Lung Stem Cell Behavior

Ahmed El-Hashash

Lung Stem Cell Behavior

 Springer

Ahmed El-Hashash
The University of Edinburgh-Zhejiang International campus
(UoE-ZJU Institute), and Centre of Stem Cell and Regenerative Medicine
Schools of Medicine & Basic Medicine, Zhejiang University
Haining, Zhejiang, China

ISBN 978-3-030-07007-6   ISBN 978-3-319-95279-6   (eBook)
https://doi.org/10.1007/978-3-319-95279-6

Printed on acid-free paper

This Springer imprint is published by the registered company Springer Nature Switzerland AG.
The registered company address is: Gewerbestrasse 11, 6330 Cham, Switzerland

This book is dedicated to my parents, kids HOOR and NOOR, wife, and new baby LIEN.

Foreword

It is with great pleasure that I pen this foreword to *Lung Stem Cell Behavior*. The field of stem cell biology and behavior is moving extremely rapidly as the concept and potential practical applications have moved from theoretical concepts into human clinical trials with often outstanding results and has thus entered the mainstream. Despite this worldwide intensity and diversity of endeavor, there remain a smaller number of book volumes that are focused almost entirely on the behavior of stem cells, and this volume is an example of this kind of books that also brings the novel scientific findings of most of the leaders of this research field together.

The concept of stem and progenitor cells has been known for a long time. However, it was the progress toward embryonic stem cells which truly leads the field of stem cell research and related field such as regenerative medicine. Embryonic stem (ES) of the mouse cells originally came from many research studies that aimed at identification of the mechanisms that control and progress of embryonic differentiation. Despite being magnificent, the cell differentiation in culture was overshadowed experimentally by their use as a vector to the germline and hence as a vehicle for experimental mammalian genetics. These studies led to research on targeted mutation in up to one third of gene loci and an ongoing international program to provide mutations in every locus of the mouse gene. A positive outcome of these studies has been to greatly illuminate our understanding of human genetics. In addition, promising researches that focus on discovering the equivalent human embryonic stem cells will certainly provide a universal source of a diversity of tissue-specific precursors, as a resource for tissue repair and regenerative medicine.

Progress toward understanding various aspects of stem cell behavior, including self-renewal, cellular differentiation, pluripotentiality, and the control of the balance between self-renewal and differentiation, that is, fundamental developmental biology at the cell and molecular level, now stands as a gateway to major future clinical applications. This book provides a timely, up-to-date, state-of-the-art reference to stem cell behavior in the lung.

Understanding the behavior of endogenous tissue-specific stem and progenitor cells in various organs such as the lung, which is the focus of this book, will greatly enhance the use of stem cells in tissue repair and regeneration and as a therapy for a wide range of human diseases. The rapid advances of induced pluripotent stem cell research and potential clinical applications point to the great possibilities of patient-specific ad hominem treatment. As Sir Martin Evans, who is credited with discovering embryonic stem cells and received the Nobel Prize for Physiology or Medicine in 2007, said, this patient-specific ad hominem treatment will help to open the door to personalized medicine, in which patients are stratified into different groups and therapeutics are tailored based on the individual patient's response. However, thus far the high costs associated with this technology may not allow it to be commercially viable, as Sir Evans said. Quite properly, much of this book concentrates on the fundamental developmental and cell biology of lung stem cell self-renewal and differentiation from which the solid clinical applications will arise.

Stem cell biology is a knowledge-based field which has come a long way since its initial conception. Although there is still a long way to go, most researchers are excited about its future impact on improving human health and healthcare. We now understand many of the major principles of cell self-renewal and differentiation that were unknown several decades ago. There are still areas like developmental niches, cell-cell and cell-growth factor interactions, and more about the epigenetic programming which maintains the stability of the differentiated state of stem cells that need to be further elucidated. These are pertinent areas that this book covers.

Prof. Alexzander A. A. Asea is currently Professor and Consultant Immunologist at the University of Texas MD Anderson Cancer Center (Houston, USA). He obtained his PhD from the University of Gothenburg (Sweden) where his studies formed the basis for clinical trials of combined histamine and interleukin-2, a drug now known as Ceplene™, currently prescribed for patients with metastatic melanoma and high-risk acute myelogenous leukemia (AML). Prof. Asea is a highly innovative and accomplished world-renowned research scientist and visionary executive academic leader with exceptional executive leadership experience spearheading strategic planning, research, training, education, and commercialization initiatives. He has received numerous honors and awards and has received grant funding from the federal government, industry, private foundations, and local community groups. Prof. Asea currently has 5 pending patents, over 255 scientific publications, books, reviews, news headliners, and editorials in a wide range of medical disciplines including stem cell biotherapeutics, cancer, diabetes, obesity, neurosciences, cardiovascular disease, exercise immunophysiology, aging, nanotechnology, thermal therapy, medicinal plants, and biomarker discovery.

University of Texas MD Anderson Cancer Center   Alexzander A. A. Asea
Houston, TX, USA

Foreword

It is with great pleasure that I compose this foreword to *Lung Stem Cell Behavior*. Stem cells are found in almost all organisms from the early stages of development to the end of life. There are several types of stem cells that have been reported. These different stem cell types may prove useful for medical research; however, each of the different types has both promise and limitations.

Stem cell is a fast-growing field of research. Sir Martin Evans, who is credited with discovering embryonic stem cells, received the Nobel Prize for Physiology or Medicine in 2007; Shinya Yamanaka, who discovered how to reprogram differentiated cells into induced pluripotent stem (iPS) cells, won the Nobel Prize in 2012 for the achievement. Many more researches have been done, and a lot of discoveries have been published. Much has been learned in a relatively short time on the behavior of lung stem and progenitor cells and their potential applications in tissue repair and regeneration as well as in the clinic. This book covers the latest advances in lung stem cell behavior research, in comparison with other important organs.

Regenerative medicine is a branch of translational research in tissue engineering and molecular biology, which deals with the "process of replacing, engineering or regenerating human cells, tissues or organs to restore or establish normal function." This field holds the promise of engineering damaged tissues and organs via stimulating the body's own repair mechanisms to functionally heal previously irreparable tissues or organs. Applying recent discovers in stem cell biology and regenerative medicine, in the lung, for example, to the betterment of human diseases has brought forth much hope but continues to present many challenges. The hope for cures has motivated different states and countries worldwide to invest in stem cell and regenerative medicine research.

In the last decades, major advances have occurred in both stem cell and regenerative medicine fields, with several new classes of stem cells being described. Thus, if one were to do a PubMed search with the term "stem cells," it would yield at least 200,000 entries related to scientific research articles published in peer-reviewed journals covering the last five decades. Notably, a similar search for the prior five decades yields only a total of less than a hundred references. It is obvious that the stem cell research field is fast growing and that this rapid growth has major advantages and implications for the future of stem cell knowledge and clinical translation. The characterization of induced pluripotent stem (iPS) cells, for example, brought many more avenues of research and discovery. Many clinical trials were initiated and conducted with multiple different stem cell types, including those derived from embryonic, cord, bone marrow, amnion, and fat sources, and manufacturing processes are now in place for a wide range of stem cell types. Thus, in 2012, the first paper showing results of two patients treated with human embryonic stem cells (hESCs), which were first isolated just 14 years ago, in 1998, was published.

Many studies have been carried out on endogenous stem cells in different organs to achieve a greater understanding of tissue turnover and responses to injury. A wide range of these studies is focusing on how we can harness the power of endogenous stem cells as a source for regenerative medicine. Currently, successful clinical application has been achieved for some organs, including the hematopoietic system. However, difficulties in identification, isolation, and expansion of many of these cell types ex vivo in other organs such as the lung have limited their widespread application.

Progress toward the understanding of various aspects of stem cell behavior such as self-renewal, cellular differentiation, pluripotency, and the control of the balance between self-renewal and differentiation, that is, fundamental developmental biology at the cell and molecular level, now stands as a gateway to major future clinical applications. This book provides a timely, up-to-date, and state-of-the-art reference in the lung, and much of it properly concentrates on the fundamental cell and developmental biology of lung stem cell self-renewal and differentiation from which the solid applications will arise.

Finally, stem cells have recently attracted much attention largely because of their potential therapeutic use in regenerative medicine and tissue repair and for developing therapies for a wide range of diseases such as cancer to eliminate cancer stem cells. Understanding the basic molecular, cellular, and genetic mechanisms that regulate stem behavior such as cell proliferation/self-renewal and differentiation, which is the subject of this book on the lung, is a very hot topic in stem cell biology and medicine and developmental biology and will lead to harness the ability of these cells in tissue repair and regeneration after injury.

Dr. Li Dak Sum & Yip Yio Chin Center
for Stem Cell and Regenerative Medicine,
Schools of Medicine & Basic Medicine, Zhejiang University   Junfeng Ji
Hangzhou, Zhejiang, China

Preface

The development of all humans begins after the union of male and female gametes or germ cells during fertilization or conception. The fertilized egg or zygote is a large diploid cell that is the beginning, or primordium, of a human being. This fertilized egg undergoes rounds after rounds of both highly organized and tightly controlled cell divisions until it comprises many billions of stem and lineage-specific cells that have self-renewal and self-repairing capabilities and form the human body. These processes are studied in a branch of science called developmental biology that explores how organisms develop and progress. As a stem cell and developmental biologist, who have investigated the mechanisms of organogenesis in a wide variety of tissues and organs such as the placenta, kidney, lung, and neural crest cells, it becomes clear to me that if we can understand these normal and fundamental mechanisms of developmental biology, then correcting abnormalities caused by congenital defects, repairing the injured tissues and even generation of functional whole organs from stem cells should be theoretically achievable. The stem cell field has grown very rapidly over the past decade and continues to be one of the most exciting aspects of biomedical research.

Stem cell research can be traced back for more than 20 years when scientists first isolated embryonic stem cells from mouse blastocysts, and a research article announcing the discovery of human embryonic stem cells emerged in 1998. Stem cell research is a fast-growing field that has rapidly expanded as new research and experience broaden our knowledge about different aspects of stem cell biology and applications. In the past decade, the stem cell field has grown very rapidly and continues to be one of the most exciting aspects of biomedical research.

Both embryonic and adult stem cells are currently a remarkably fast-growing field of research, with an astonishing annual growth rate of 77% since 2008. The volume of research output, and thus publication, has therefore increased significantly in all areas of stem cell research. By now, the first functioning whole organ, thymus, has been generated in the laboratory, and the first in vitro fertilized human baby girl has children of her own. Research is currently underway in different laboratories worldwide to generate other functioning whole organs such as the intestine, kidney, and other human body organs.

Embryonic stem cells (ESCs) were isolated from mouse blastocysts by scientists in 1981, while human ESCs were first reported in 1998. Currently, adult-derived stem cells (ASCs) are also a favorite subject of intensive research investigations. Recently, ESCs are almost routine in the face of more advances in the medical field. More recent advances show the possibility to turn fully differentiated cells *back* into a more embryonic-like state of induced pluripotency. This occurs by means of as few as four factors and represents a major scientific discovery. Moreover, it has been recently shown that several classes of stem-like cells, which are originating from different mesenchymal compartments of the body such as the amnion, marrow, amniotic fluid, and adipose, exert promising therapeutic effects in some inflammatory and fibrotic diseases. In addition, neural stem cells can be programmed to selectively travel and attack inaccessible brain tumors. Furthermore, the recent identification of endophenotypes, or latent risk factors, for certain types of aggressive cancers may eventually lead to designing novel strategies for cancer treatments. Together, these recent discoveries could identify the next generation of treatments emerging from our scientific discoveries.

Scientists worldwide are applying new stem cell discoveries to the betterment of human diseases, which have brought forth much hope for better human life. The branch of translational research in the tissue engineering and molecular biology, which takes advantage of rapid progress in our understanding of stem cell biology during development and adulthood, is called regenerative medicine. The hope for cures of different diseases has prompted different countries worldwide to invest in stem cell research and regenerative medicine.

The USA, for example, plays a critical role in stem cell research, like a lot of other countries in the world. Many countries in Europe, Australia, Japan, China, and other Asian countries, in addition to Canada and Brazil, have leading centers for stem cell research and regenerative medicine. These research centers have significantly expanded the scope of stem cell researches and their applications in the treatment of different human diseases.

New insights have been added to the identification and characterization of endogenous tissue-specific stem and progenitor cells in the lung over the last few years. The exploration of endogenous lung stem and progenitor cells holds promise for advancing our understanding of the biology of lung repair and regeneration mechanisms after injury. This will also help in the future use of stem cell therapies for the development of regenerative medicine approaches for the treatment of different lung diseases.

This book brings together several topics that are related to lung stem cell behavior, biology, and development as well as stem cell applications in lung repair and regeneration. We discuss recent advances in the identification and characterization of the main types of lung stem and progenitor cell populations. In addition, we describe recent research progresses and accumulated information regarding the behavior, development and function of various lung stem and progenitor cells, and factors that control their repair and regeneration after injury as well as the molecular and cellular mechanisms that regulate lung stem/progenitor cell behavior during development and repair and regeneration. We also briefly describe stem cell behaviors and their regulatory molecular mechanisms in other systems or organs to explore their similarity and difference with lung stem cells.

This book contains a review for a global collection of recent monograph essays from a wide range of research scientists who are investigating different aspects of lung stem cell biology at various research institutes and countries. It describes and discusses exciting progresses in basic stem cell behavior and biology and regenerative medicine, including the potential applications of stem cells in lung repair and regeneration as well as in some lung diseases.

Although we could not hope to be comprehensive in the coverage of stem cells of other tissues, organs, or systems, our main goal in compiling this book was to bring together a selection of the current progress in understanding the behavior and development of endogenous tissue-specific stem and progenitor cells in the lung, as well as the potential applications of stem cells in lung repair and regeneration after injury or diseases. In preparing this book, we aimed at making it accessible not only to those working in stem and progenitor cell behavior and biology fields but also to non-experts with a broad interest in stem cells and regenerative medicine in human health. Our hope is that this book will be of value to all concerned with stem cell biology, development, and application in medicine.

Haining, Zhejiang, China   Ahmed El-Hashash

Acknowledgments

Eiman Abdel Meguid, MD

Senior Lecturer, Centre for Biomedical, Sciences Education, School of Medicine, Dentistry and Biomedical Sciences Queen's University, Belfast, Ireland, UK

Esam I. Agamy, PhD

Professor, College of Medicine, University of Sharjah, Sharjah, UAE

Wadah AlHassan, BSc

California State Polytechnic University, Pomona, 3801 West Temple Avenue Pomona, CA 91768, USA

Alexzander A. A. Asea, PhD

Visiting Professor and Consultant Immunologist, Center for Radiation Oncology Research, Department of Experimental Radiation Oncology, The University of Texas MD Anderson Cancer Center, 1515 Holcombe Boulevard, Houston, TX 77030, USA

Karen Ek, BSc

California State University San Bernardino, 5500 University Pkwy, San Bernardino, CA 92407, USA

Magdy Elhefnawy, MD

President, Gharbia Medical Syndicate, Tanta University Medical School, Tanta, Egypt

Reda Elnagar, BSc

Senior Lecturer, Basuin high School of Girls, Basuin City, Gharbia Governorate, Egypt

Haifen Huang, BSc

California State Polytechnic University, Pomona, 3801 West Temple Avenue, Pomona, CA 91768, USA

Junfeng Ji, PhD

Professor of Stem Cells and Regenerative Medicine, Chairman, Dr.Li Dak Sum & Yip Yio Chin Centre for Stem Cell and Regenerative Medicine, School of Medicine, Zhejiang University, 866 Yuhangtang Road, Hangzhou, Zhejiang 310058, China

Susan J. Kimber, PhD

Professor of Stem Cell and Developmental Biology, The University of Manchester, Manchester, England, UK

John Ku, BSc

California State Polytechnic University, Pomona, 3801 West Temple Avenue, Pomona, CA 91768, USA

Linrong Lu, PhD

Professor of Immunology, School of Basic Medical Sciences, Zhejiang University, 866 Yuhangtang Road, Hangzhou, Zhejiang 310058, China

Karol Lu, BSc

University of Southern California, University Park, Los Angeles, CA 90089, USA

Gamal Madkour, PhD

Professor, Tanta University School of Science, Tanta University, Tanta, Egypt

Moustafa Mahmoud, PhD

Professor, Tanta University School of Medicine, Tanta University, Tanta, Egypt

Contents

About the Author

Ahmed El-Hashash has completed his PhD from Manchester University, UK. He is a fellow of the California Institute for Regenerative Medicine (CIRM) and New York University Medical School (MSSM), USA. Prof. Ahmed Hashash worked as a Senior Biomedical Research Scientist at Mount Sinai School of Medicine of New York University and Children's Hospital Los Angeles. He was Assistant Professor and Principal Investigator of Stem Cell & Regenerative Medicine at Keck School of Medicine and Ostrow School of Dentistry of the University of Southern California, USA. Prof. Hashash has joined The University of Edinburgh-Zhejiang International Campus (UoE-ZJ) as Tenure-Track Associate Professor and Senior Principal Investigator of Biomedicine, Stem Cell & Regenerative Medicine. He is also adjunct Professor and Senior Principle Investigator at the School of Basic Medical Science and School of Medicine, Zhejiang University. Prof. Hashash has several break-through discoveries in genes/enzymes that control stem cell behavior and regenerative medicine. He has published more than 33 papers (and 20 conference abstracts) in reputed international journals and serving as an editorial board member of repute. Prof. El-Hashash acts as a Discussion Leader at the prestigious Gordon Research Seminar/Conference in the USA and a Peer Reviewer/International Extramural Review for The Medical Research Council (MRC) grant applications, London, UK. He is invited to speak at several international conferences in the USA, Spain, Greece, Egypt, Japan, France, and China. He is the editor or author of several books on stem cell and regenerative medicine.

List of Figures

Introduction

In life, complex living organisms are faced with multiple major challenges, including the generation and maintenance of a wide range of diverse cell types, such as simple epithelia, muscles, bones, neurons, secretory glands, and motile immune cells. Complex sets of molecular and functional mechanisms as well as signaling pathways are required to control nearly every aspect of cell growth, development, and function. These mechanisms and pathways include many genetic and epigenetic programmings, cell-cell communications, cytoskeleton dynamics, cell adhesions, and protein traffickings (Seita and Weissman 2010; Kohwi and Doe 2013; Clevers et al. 2014; Drummond and Prehoda 2016).

The term "stem cell" describes the self-renewing, primitive, undifferentiated, and multipotent source of multiple cell lineages. While such cells are critical for growth and development through childhood, pools of adult stem cells are hypothesized to be the source of the often somewhat limited tissue regeneration and repair in adults. In contrast to embryonic stem cells and tumor cells, adult human stem cells show decreasing telomere length with increasing age (Warburton et al. 2008).

In different tissue types, stem cells are undifferentiated or "blank" cells that can characteristically undergo self-renewal and then differentiation into a wide range of cell types that perform different biological functions. Stem cells are, therefore, multipotent and act as a source of multiple cell lineages, a function that is essential for the formation and development of human body as well as for the repair and regeneration of different body's tissues after injury in childhood and in adult life (El-Hashash 2016a, b). The ability of stem cells to undergo self-renewal is also critical for preserving the stem cell pool in each organ, while their differentiation into many specialized cell types is essential for body functions and occurs in response to proper signals from other cells. Therefore, stem cells have major functions in tissue formation, development, repair and regeneration, and in the healthy cell turn over and tissue homeostasis (reviewed in El-Hashash 2013, 2016a, b).

A delicate balance between the self-renewing and differentiation of many tissue-specific stem and progenitor cells is essential to generate a wide range of cellular diversity during development and both tissue repair and regeneration after injury. This fine balance is also essential to maintain adult tissue homeostasis. This balance can be achieved through controlled asymmetric cell divisions. Notably, dysregulated asymmetric cell divisions, and consequently, disruption of the fine balance between the self-renewal and differentiation of stem/progenitor cells, may lead to a severe premature reduction in the number of stem and progenitor cells and defects in tissue development. This may also result in the development of severe human diseases such as cancer in different types of tissues. *However*, it is still unknown the causal relationship between dysregulated asymmetrical mode of cell divisions and the development of cancer (Gómez-López et al. 2014).

Therefore, the mode of division of stem cells is essential not only for their maintenance but also for their expansion during tissue development, morphogenesis, and repair and regeneration after injury. The cell division may occur either through symmetric or asymmetric mode. The symmetric division leads to the formation of cells that have the same fate (i.e., can self-renew/proliferate or differentiate), while the asymmetrical mode of cell division can produce two daughters with different cell fates. Thus, in the asymmetric cell division, one daughter cell can normally retain its identity as a stem cell, while the other one differentiates into a specific cell lineage. This asymmetric mode of cell division is important to maintain the cellular diversity of multicellular organisms (Peng and Axelrod 2012; Gómez-López et al. 2014; Dewey et al. 2015).

During the development, several molecular cues within the cell and in its surrounding environment function to direct stem cells to have either symmetrical or asymmetrical cell divisions. These two types of cell divisions can be identified and distinguished by observing the spindle fiber orientation and the differential inheritance of several polarity proteins, including Numb and atypical protein kinase Cζ (aPKCζ) that are critical for the cell fate determination (Huttner and Kosodo 2005; Morrison and Kimble 2006; El-Hashash and Warburton 2011, 2012; El-Hashash et al. 2011c; Drummond and Prehoda 2016). The asymmetric cell division may occur in response to either intrinsic or extrinsic fate determinant molecules. Thus, daughter cells that are placed in two biologically different microenvironments may acquire completely different cell fates due primarily to differences in the extrinsic fate determinants in their respective microenvironment. In contrast, the intrinsic and cytoplasmic cell fate determinants, including the Notch signaling inhibitor Numb, can be asymmetrically localized within the mitotically dividing epithelial cell. Then, Numb protein differentially segregates into daughter cells that will acquire different fates (reviewed by Yamashita 2009).

The ability of different types of stem and progenitor cells to determine and control their cell fate decisions is crucial during the development of animals and humans. Proper execution of cell fate specifications and decisions ensures tissue development and morphogenesis as well as homeostasis throughout adulthood. Conversely, any defect in the molecular and cellular mechanisms that regulate the cell fate specification and decision may lead to the development of a wide range of diseases. Much *accumulated* recent *research has* identified many evolutionarily conserved protein complexes, cell fate determinant proteins, and both molecular mechanisms and signaling pathways that regulate different aspects of cell fate decisions across a wide variety of tissues. The functions of cell fate determinant molecules and signals in maintaining a proper cell fate determination process are critical during mitosis in different tissue types (Dewey et al. 2015).

Until most recently, the presence of specific self-renewing cells has been uncovered in the lung. In addition, little is known about whether a single stem cell can effectively generate a wide range of different cell types (more than 40 cell types), which are required for proper formation and function of the lung. In the adult mouse airway, developmental and stem cell biologists have identified at least five putative epithelial stem or progenitor cell niches (Lu et al. 2008; Green et al. 2013). They have also identified both airway-specific smooth muscle stem cells and endothelial stem cells in the lung vasculature (Lu et al. 2008; Green et al. 2013). Some circulating stem and progenitor cells may reside in the lung and can, therefore, be considered as a source of lung stem cells (Lu et al. 2008; Green et al. 2013).

Many epithelial tissues, including the skin and gastrointestinal tract, have a characteristic rapid regeneration capacity, while the lung shows a remarkably slow turn over. This slow turn over and the lack of both specific makers and clonality assays for lung stem/progenitor cells have remarkably hampered the localization, identification, and isolation of putative and specific stem/progenitor cells in the lung, compared to other organs (reviewed by El-Hashash 2013; Ibrahim and El-Hashash 2015).

This book is meant to provide an introduction to human respiratory system (Chap. 1) and to discuss recent research investigations on the identification of different types of lung stem and progenitor cells (Chap. 2) and both signals and molecular mechanisms that control stem/progenitor cell behavior in the lung (Chap. 3) compared to other well-studied systems (Chap. 4). This book also aims to cover and discuss recent studies on the mode of cell division and its regulatory mechanisms in mammalian lung stem and progenitor cells (Chap. 5) and lung stem cell plasticity (Chap. 6) as well as the functions and applications of stem cells in both lung repair and regeneration (Chap. 7) and modelling lung development and diseases (Chap. 8). In addition, we comprehensively discuss rapidly growing research data on the molecular mechanisms and signal pathways controlling the cell polarity and asymmetric cell division, as well as balancing self-renewal with differentiation in lung stem and progenitor cells during lung development and repair and regeneration. We hope that this book will provide a better understanding of the molecular and cellular mechanisms that regulate the behavior of lung stem/progenitor cells that can help to identify new targets for preventing and rescuing infants and children from lethal lung diseases and for the regeneration of damaged lungs.

Chapter 1
Brief Overview of the Human Respiratory System Structure and Development

Abstract Human respiratory system is one of the biological systems that involves specific tissues and organs that are responsible for the breathing process, the air quality, and the gas exchange that are essential processes for human life. The respiratory tract characteristically and intensively interacts with the environment. In this chapter, we will overview the structure of the human respiratory system. We will also briefly describe the developmental stages of human respiratory system.

Keywords The respiratory system · Larynx · Trachea · Bronchi · Airways · Lungs · Branching morphogenesis

1.1 Brief Overview of the Structure of Human Respiratory System

The human respiratory system has two divisions: the upper division of the respiratory tract, which contains the nose and pharynx, and the lower division of the respiratory tract that consists of the larynx, trachea, right and left bronchi, and lungs. The bronchial tree consists of different parts such as the primary (main) bronchus, secondary (lobar) bronchus for each lobe, tertiary (segmental) bronchus for each segment, conducting (lobular) bronchiole (1 mm; no cartilage), terminal bronchiole, respiratory bronchiole, and alveolar duct as well as the alveoli (recently reviewed in Rizzo 2016; Marieb and Keller 2017).

Air first enters the respiratory system through the nose or mouth and then reaches the pharynx via the nasopharynx. When it reaches the glottis, air passes to the trachea and then into both the right and left bronchi and their branched small bronchioles and terminal alveoli sacs. There are almost 300 million alveoli that provide at least 160 m^2 of surface area for gas exchange in two adult lungs (Rizzo 2016; Marieb and Keller 2017).

The pharynx consists of three parts: nasopharynx, oropharynx, and laryngopharynx. The nasopharynx localizes to the posterior of the nasal cavity and above the soft palate. The auditory tube opening exists on lateral wall of the nasopharynx. This part of pharynx functions as a passage for air. The oropharynx is the oral part

© Springer International Publishing AG, part of Springer Nature 2018
A. El-Hashash, *Lung Stem Cell Behavior*,
https://doi.org/10.1007/978-3-319-95279-6_1

of pharynx and lies posterior to oral cavity, extending from below soft palate to the level of vertebra C3. On the lateral wall of the oropharynx are the palatine tonsils. The oropharynx is the part of the pharynx that is common for air and food. The laryngopharynx is the part of the pharynx that extends from the oropharynx above and continues as esophagus. It lies behind the opening into the larynx and extends to the level of C6. This part of the pharynx is also common for air and food (Rizzo 2016; Marieb and Keller 2017).

The human trachea is normally 13 cm long and 2.5 cm in diameter. It has a fibro-elastic wall made of U-shaped bars of hyaline cartilage, which keep the lumen patent. The trachealis muscle is a smooth muscle that connects the posterior ends of the tracheal cartilages. The trachea commences in the neck below the cricoid cartilage (the sixth cervical vertebra) and divides into both the right and left bronchi at the sternal angle. This bifurcation is called the carina. During deep inspiration, the carina descends to the level of the sixth thoracic vertebra (Rizzo 2016; Marieb and Keller 2017).

Lungs are half cone in shape, with a base, apex, two surfaces (costal and mediastinal), and three borders (anterior, inferior, and posterior). The lung base sits on the diaphragm, while the lung apex projects above the first rib and into the root of the neck. The hilum is where structures enter and leave. The heart and major vessels indent the mediastinal surface of lung.

The lungs are surrounded by pleura. While the parietal pleura acts to cover some parts of the respiratory system such as the diaphragm, thoracic wall, and mediastinum, the visceral pleura functions as a cover for the outer surfaces of the lungs and dips into its fissures. The pleural cavity localizes between these two layers and is filled with pleural fluid secreted by the pleural membranes. The pleural cavity functions to provide lubrication and, thereby, prevents friction during breathing. The parietal pleura consists of the cervical pleura, the costal pleura, the diaphragmatic pleura, and the diaphragmatic pleura. In addition, the parietal pleura is reflected from the mediastinum to the lung to form the visceral pleura. Such an arrangement creates recesses in the pleural cavity between the layers of the pleural reflections (Rizzo 2016; Marieb and Keller 2017).

1.2 Brief Overview of the Development of Human Respiratory System

The development of human respiratory system occurs via a series of consecutive stages. Division of the foregut tube into two parts, the developing respiratory endoderm anteriorly and the esophagus posteriorly, is the early stage of the respiratory system development. Intensive branching morphogenesis during the second stage leads to the formation of the proximal conducting airways, followed by a remarkable distal septation that is able to generate the gas exchange units, which are the alveoli. Normal lung formation and development require a simultaneous and proper formation of the closely related pulmonary and bronchial vascular systems (Warburton et al. 2010; Rankin and Zorn 2014; Schittny 2017; Chen and Zosky 2017).

The laryngotracheal groove gives rise to the lung after its invagination into the surrounding splanchnic mesenchyme. The human lung is derived from the anterior surface of the growing foregut at 5 weeks of gestation and continues to grow for several months after birth. Different parts of the respiratory tract, including the larynx, trachea bronchi, and lungs, are formed from the growing respiratory bud that starts to appear ventral to the caudal portion of the foregut during the fourth week of human gestation (Warburton et al. 2010; Rankin and Zorn 2014; Schittny 2017; Chen and Zosky 2017).

At the end of the fourth week of gestation, a pair of primary human bronchial buds, called left bud and right bud, evaginate from the laryngotracheal groove. The right bud gives three secondary bronchial buds, while the left bud gives two secondary bronchial buds during the fifth week of human gestation. The developing secondary bronchial buds form the lung lobes and undergo more branching to generate the tertiary buds over the following week of gestation. With continuous development and growth, the major elements that form the lungs appear, except the alveoli between the sixth week and the sixteenth week of gestation. From week 16 to week 26 of gestation, a prominent increase in both the size of the bronchi and lung tissue vasculature occurs, which is associated a continuous development of both the bronchioles and alveolar ducts. In human, alveolarization starts around 20 weeks of gestation. It continues postnatally till 7 years of age (Warburton et al. 2010; Schittny 2017).

There are 23 generations of airway branching in the human lung, with the first 16 generations characterized by a stereotypic branching that is almost completed by 16 weeks of gestation. The other seven branching generations are almost completed at the 20–24 weeks of gestation. During lung development, both the characteristic type I alveolar epithelial cells, in which gas exchange occurs, and the type II alveolar epithelial cells that produce pulmonary surfactant are formed. The surfactants are phospholipids that function to reduce the surface tension at the air-alveolar surface. These phospholipids, therefore, help the terminal saccules to expand and function to prevent the collapse and sticking of the alveoli (Warburton et al. 2010; Schittny 2017).

Chapter 2
Diversity of Lung Stem and Progenitor Cell Types

Abstract Stem cells are multipotent/pluripotent source of a wide range of cell lineages that are crucial for tissue development, repair, and regeneration. They are undifferentiated cells that can characteristically self-renew and develop into different cell types that perform various functions in the human body. Stem cells play critical roles in body development and growth through childhood. Recent data indicate that pools of adult stem cells are also important in some examples of the repair and regeneration in adult tissues. Stem cells can self-renew to maintain the tissue-specific pool of stem cells and differentiate into various functionally specialized cell types. Considerable data have accumulated in recent years on the characterization, isolation, and function of stem/progenitor cells in the lung. In this book chapter, we describe accumulated data on the localization, characterization, biology, and function of different lung stem cell populations such as alveolar, bronchial, and tracheal epithelial cells, as well as endogenous mesenchymal stem/progenitor cells.

Keywords Lung · Alveolar epithelium · Stem cells · Bronchial · Tracheal epithelial cells · Mesenchymal cells

There is increasingly accumulated knowledge about specific types of self-renewing cells in the lung. It seems unlikely that a single cell (e.g., a stem cell) is sufficient for generating over 40 distinct types of cells that are required for the functions of the lung. In the adult mouse airway, at least five epithelial stem cells niche, beside other related stem cells such as airway smooth muscle and endothelial stem cells (Liu et al. 2007, 2008; Green et al. 2013). In addition, some circulating stem cell types can reside in the lung tissues. Furthermore, the lung has a remarkably slow turn over compared to other tissues of epithelial origin such as the skin and intestine that are characterized by their rapid regeneration. This remarkably slow turn over hitherto has impeded the definition and characterization of new lung stem and progenitor cell populations. In addition, there are relatively insufficient markers and clonality assays to identify and isolate lung-specific stem/progenitor cells (El-Hashash 2013;

© Springer International Publishing AG, part of Springer Nature 2018
A. El-Hashash, *Lung Stem Cell Behavior*,
https://doi.org/10.1007/978-3-319-95279-6_2

Ibrahim and El-Hashash 2015). In the following sections, we review new information that has recently been accumulated on lung-specific stem and progenitor cells,
including their classification, development, function, and repair/regeneration
capacities.

2.1 Alveolar Epithelial Stem and Progenitor Cells

2.1.1 Localization and Characterization

Many epithelial-derived stem/progenitor cells that are in an undifferentiated state,
multipotent, and express characteristic genetic markers exist at the distal tips of
epithelial branching tubules during the pseudo-glandular stage of lung formation
(Rawlins et al. 2009a; Figs. 2.1 and 2.2). In addition, different types of putative
endogenous epithelial stem/progenitor cells exist at the upper airway basal layer,
while the alveolar epithelium localized to the bronchoalveolar junction, and close to
or within the rests of pulmonary neuroendocrine cells in the adult lung (Fig. 2.1;
Reynolds et al. 2000; Engelhardt 2001; Giangreco et al. 2002; Giangreco et al.
2004; Reddy et al. 2004; Reynolds et al. 2004; Kim et al. 2005; Rawlins and Hogan
2006; Jiang and Li 2009; Rock and Hogan 2011).

   The distal lung epithelium possesses multipotent progenitors that contribute
descendants to both the bronchi and alveoli during lung early formation and development (Rawlins et al. 2009a, b). The descendants of distal stem/progenitor cells
are left behind in the lung epithelial stalk, while proliferative stem cells continue to
reside in the budding tips during epithelial branching morphogenesis. There are
several evidences that support this model, including the unique expression pattern
of the transcription factors: inhibitor of differentiation 2 (Id2), SRY-box containing
gene 9 (Sox 9), and v-myc myelomatosis viral related oncogene (N-myc), as well as
ets variant gene 5 (Etv5/ERM), are expressed at high levels in the distal epithelial
cells of the developing lung (Fig. 2.2). In addition, these distal lung epithelial cells
are regulated by several major signaling pathways such as BMP (bone morphogenetic protein), SHH (sonic hedgehog), and FGF (fibroblast growth factor), as well
as Wnt protein (Bellusci et al. 1996; Liu et al. 2002; Shu et al. 2005) that will be
discussed in detail in Chap. 3. A lot of these regulatory signaling pathways and
characteristic gene markers, including FGFR signaling and both Sox9 and etv5
genes, are also important for the stem/progenitor cell development in other
endodermal-derived organs, including the pancreas (Zhou et al. 2007; Seymour
et al. 2007; Rock and Hogan 2011). Moreover, epithelial cells of the distal lung have
a unique kinetics of cell cycle that is different from other lung epithelial cells
(Okubo et al. 2005). In addition, a high proportion of these distal lung epithelial
cells can rapidly incorporate the thymidine analog bromodeoxyuridine (BrdU)
within 1 h. (Okubo et al. 2005).

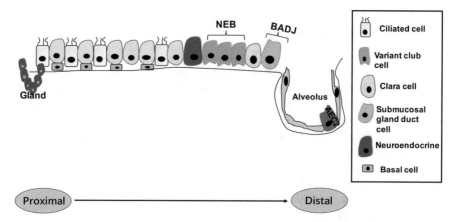

Fig. 2.1 Candidate stem/progenitor cells in the lung epithelium. Schematic depiction of proposed epithelial stem and progenitor cells and their niches in the proximal conducting airways and distal alveoli. *Abbreviations*: AEC2 type 2 alveolar epithelial cell; BADJ bronchoalveolar duct junction; Gland submucosal gland duct; NEB neuroepithelial body

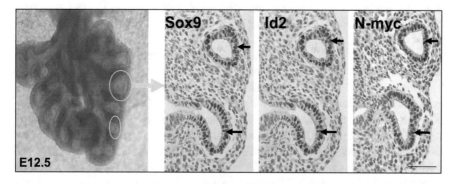

Fig. 2.2 Lung epithelial stem and progenitor cell markers during early embryogenesis. External appearance (left hand panel) and immunoperoxidase staining of day (E) 12.5 embryonic lung show that lung epithelial stem/progenitor cells localized to distal epithelial tips (yellow circles) and expressed Sox9, Id2, and N-myc proteins (arrows). Scale bar: 100 μm

The alveolar epithelium contains alveolar type I (AT1) cells that are flat and consist of most of the surface area for the gas exchange (95%). It also contains less than five percent of alveolar type II (AT2) cuboidal cells. In addition, a subgroup of stem cells is detected within the AT2 cells and is widely distributed in the terminal bronchiole and alveolus, as well as bronchoalveolar duct junction (BADJ). There is a clear difference in both the differentiation potential and numerical preponderance between AT2 cells and bronchoalveolar stem cells (BASCs; Li et al. 2015; Sun et al. 2013). Moreover, BASCs can be generated by some AT2 cells in murine lungs (Driscoll et al. 2000).

2.1.2   The Repair and Regeneration Capacities and Mechanisms

Many epithelial stem cell populations may serve as a "ready reserve" for repairing alveolar cells that are damaged after injury. The failure of endogenous stem cells to do this function may lead to the decline of tissue repair and/or regeneration during aging. For example, telomerase is marker of lung stem and progenitor cells, and its expression is strongly and broadly upregulated in the alveolar cells in the recovery phase after the acute oxygen injury (Driscoll et al. 2000). Without the telomerase expression, the repopulation of damaged alveoli and resistance to injury are both severally changed, suggesting that telomerase is likely important for the activity of alveolar stem/progenitor cells (Jackson et al. 2011). In addition, alveolar epithelial cells show DNA damages and other types of cell damages, including the depletion of glutathione, failure of the mitochondria, and cell death after the acute oxygen injury (Buckley et al. 1998; Roper et al. 2004).

Recent advances in the development of more effective lineage-tracing systems have led to improving our understanding of the origin, development, behavior, and biology of lung-specific stem and progenitor cells that also help in identifying their roles in lung repair and regeneration after injury (Ibrahim and El-Hashash 2015; Elshahawy et al. 2016; Chen and Fine 2016). In addition, the important roles of different types of pulmonary stem and progenitor cells in the repair of injured lung tissues and recovery of homeostasis are well investigated through lineage tracing and other research techniques (Liu et al. 2007, 2008; McQualter et al. 2010).

There are evidences that some alternative populations of stem and progenitor cells exist in the lung epithelium and play important role in the reconstitutions of alveolar epithelial cells after injury (Chapman et al. 2011). These alternative populations of stem/progenitor cells at the distal epithelium, therefore, have reparative functions and properties and express markers of the basal cells such as KRT5 and TRP63, as well as can generate differentiated bronchiolar and alveolar epithelium after injury (Zuo et al. 2014). Moreover, the reconstitution of both distal airway club cells and AEC2s can occur by recently identified lineage-negative epithelial progenitors after influenza viral infection in living mice (Vaughan et al. 2015). Remarkably, the suppression of Notch signaling activity is essential for differentiating lineage-negative epithelial progenitor cells to AEC2s, while the persistence of active Notch signaling can lead to the formation of cysts that are reminiscent of honey combing in lung fibrosis in humans (Vaughan et al. 2015).

Furthermore, several changes were reported in AT1 cells, including cell injury, or necrosis/apoptosis when the acute lung injury (ALI) occurs that can also case hypertrophy of some AT2 cells (Chen et al. 2012a, b). Intestinally, these AT1 cells could be substituted in the injured tissue by some mature AT2 cells that can differentiate and produce new AT1 (Chen et al. 2012a, b). In this regard, the AT2 cell heterogeneity is arranged into three groups: the alveolar repair-focused AT cells, alveolar renewal-focused AT2 cells, and AT2 replacement-focused cells (Desai et al. 2014).

In addition, AT2 cells can differentiate to AT1 cells during the postnatal lung growth, and in response to alveolar injury of the adult lung (Evans et al. 1975).

Remarkably, bronchoalveolar stem cells (BASCs) show several characteristics of stem cells with induced proliferation after alveolar/airway injury (Kim et al. 2005). BASC cells reside near bronchiolar-alveolar junctions and express several markers such as pulmonary surfactant protein (SP-C) and Clara cells-10 kd protein (CC10) that are alveolar and airway epithelial markers, respectively, as well as stem cell antigen-1 (*Sca-1*) protein (Kim et al. 2005). They are multipotent and can self-renew and differentiate to form both alveolar and Clara cell types (Kim et al. 2005). In addition, variant Clara cells are lung-specific and can divide to repopulate the airway epithelium of the distal lung after lung injury (Hong et al. 2004). Indeed, Clara cells self-renew and can function as ciliated progenitor cells during lung continuous growth early after birth (McDowell et al. 1985; Plopper et al. 1992; Perl et al. 2005).

2.2 Tracheal and Bronchial Epithelial Stem and Progenitor Cells

2.2.1 Localization and Characterization

The trachea of the murine or human respiratory system is tightly supported by the cartilage rings and both contractile and noncontractile mesenchymal cells (Evans et al. 2001). Several types of stem and progenitor cells were recently discovered and characterized in the epithelium of the trachea, bronchi, and lungs using several animal models of lung injury.

Both Clara cells and two types of mitotic basal cells, keratin 5+/15+ double-positive and keratin 14+/5+/15+ triple-positive cells, exist in the tracheal epithelium and are important for the hemostasis of epithelial cells since they represent two independent pools of progenitor cells (Cole et al. 2010). In addition, lineage-tracing studies of the trachea and lung of adult mouse showed that cells that are keratin-14-positive are able to function as progenitor cells, while ciliated cells lack this ability (Hong et al. 2004; Rawlins et al. 2007). In addition, migrating gland duct cells can contribute to the formation of the tracheal epithelium during its regeneration after naphthalene injury (Borthwick et al. 2001).

Furthermore, Clara-like cell types are secretory and non-ciliated cells as well as express both SCGB1A1 (secretoglobin 1A1) and CCSP/CC10 (Clara cell secretory protein). They can also be identified morphologically by their apical projections (Evans et al. 1986, 2001; Rawlins et al. 2009a, b). Although Clara-like cell types show some similarities with bronchiolar Clara cells, they also have some distinct functions, structures, and other properties. Furthermore, there are evidences of interactions between ciliated cells and Clara-like cells through their adherens and gap junctions (Evans et al. 1986).

2.2.2   The Repair and Regeneration Capacities and Mechanisms

Recent studies have uncovered the cellular mechanisms regulating both the maintenance and repair of tracheal epithelial cells (Cole et al. 2010). In addition, the gland ducts represent the major source for the regeneration of airway tracheal epithelial cells after an induced injury (Borthwick et al. 2001). Moreover, the submucosal gland ducts probably contain putative stem/progenitor cells in the proximal airway (Lu et al. 2008). The function of another cell population of the airway epithelium, the submucosal gland (SMG) duct cells, in the repair of the surface epithelium (SE) and SMG tubules has been investigated after a hypoxic ischemic injury (Hegab et al. 2011). Lineage-tracing analyses and both in culture and in vivo stem cell model systems showed that the submucosal gland (SMG) duct cells divide, differentiate, and give rise to SMGs and SMG duct cells, which probably form SE in the area adjacent to the submucosal duct (Hegab et al. 2011). SMG duct cells were, therefore, proposed to function as repairing stem cells for the airway epithelium and may play major functional roles in the treatment of many lung diseases (Hegab et al. 2011).

The pseudostratified epithelial compartment contains basal cell (BC) populations that express cytokeratin 5 (Krt5) in both the murine trachea and human airways. Both lineage tracing and transgenic mouse studies showed that tracheal basal cells can act as progenitor cells during the postnatal lung growth and in the adult lung (Rock et al. 2009). In addition, the tracheal basal cells can serve as progenitor cells during the recovery of tracheal epithelial cells after a sulfur dioxide (SOX2)-induced injury (Rock et al. 2009). In addition, several clonality assay studies have further supported the importance of lung basal cells. These cells, can self-renew and differentiate into both mucus and ciliated cells in the absence of columnar or stromal epithelial cells in murine and human airways (Rock et al. 2009, 2010).

Basal cells that are positive for keratin-14 (K-14) are different from bronchial basal cells and have the ability to repair differentiated epithelial cells after injury (Hong et al. 2004). These cells can, therefore, restore the differentiated epithelial cells after injury in the trachea (Hong et al. 2004). Notably, after a depletion of Clara cells, keratin-14-positive (K-14+) basal cells can function as alternative progenitor cells (Hong et al. 2004). These cells can, therefore, contribute to cell proliferation and self-renewal in the injured bronchial epithelium (Hong et al. 2004). In addition, lineage-tracing investigations in the trachea and lung of adult mice have demonstrated that K-14-positive (K-14+) basal cells could act as progenitor cells, excluding ciliated cells (Hong et al. 2004; Rawlins et al. 2007).

An elegant lineage-tracking system has been used by Rawlins et al. (2009a, b) to circumvent many of the obstacles that have hitherto hampered study of lung stem and progenitor cells and to show that distinct distal populations of epithelial progenitors maintain the alveolar surface. Using a newly generated a "knockin" transgenic mouse model, this study showed that Clara cells play a major role in both reconstituting the epithelial tissues in the bronchioles and repair of the trachea, and these processes involve the self-renewing and differentiation of Clara cells to form ciliated cells (Rawlins et al. 2009a, b).

The Clara cells that express secretoglobin 1A1 (SCGB1A1+) have, therefore, the self-renewing and proliferation capacities in the injured trachea. However, this was not considered as the major source or key mechanism for regenerating the trachea (Rawlins et al. 2009a, b). In addition, a special subpopulation of putative bronchoalveolar stem cells (BASCs) that express both SP-C and CC10 may not have a role in postnatal lung growth and repair or adult lung homeostasis (Rawlins et al. 2009a, b).

Other cell population types were identified in the lower respiratory tract. A pluripotent cell population, for example, exists in the lower respiratory tract and is identified by their expression of airway and mesenchymal molecular markers (Giangreco et al. 2004). Another cell population is from human tracheobronchial epithelium, proliferative, and $CD45^-$ SP (Hackett et al. 2008). Notably, these cells increase in their number in asthmatic airways that suggest a role for the dysregulation of pluripotent cells in the pathogenesis of asthma (Hackett et al. 2008). Some of these SP cells also show endothelial progenitor cell potential in response to some treatments such as the hyperoxic exposure in the developing lung (Irwin et al. 2007).

2.3 Lung Mesenchymal Stem and Progenitor Cells

Lung mesenchymal stem/progenitor cells are generally not well studied, compared to epithelial stem and progenitor cells, despite many evidences that lung mesenchymal cell- derived signals play crucial roles in lung epithelial cell development and branching morphogenesis. In this section, we will describe the two major types of lung endogenous mesenchymal stem and progenitor cells: *lung vascular endothelial and smooth muscle stem and progenitor cells.*

2.3.1 Lung Vascular Endothelial Stem and Progenitor Cells

The lung mesoderm can generate different cell lineages, including the vascular endothelium and both the airway and vascular smooth muscles during embryogenesis. During early stages of lung development, the newly formed capillaries that surround the laryngotracheal groove could be easily detected by their β-galactosidase expression that is controlled by the Flk1 promoter. Flk1 promoter is, therefore, considered as the early hemangioblasts' marker. In addition, the epithelial-derived VEGF growth factor tightly regulates the differentiation of hemangioblasts into a complex network of capillaries, surrounding the developing lobar and bronchial, as well as segmental branches of the airway (Ramasamy et al. 2007; del Moral et al. 2006). Generally, the growth of both distal airway spaces and pulmonary vascular is highly coordinated and tightly controlled, which are affected by lung injury (Jakkula et al. 2000).

The microcirculation of the lung has many stem and progenitor cell types. However, little is known about these stem and progenitor cells. In addition, the mesothelial cell compartment overlies the lung and has a progenitor cell population that can differentiate to pulmonary vascular smooth muscle cells during lung forma- tion (Que et al. 2008). Moreover, vascular wall progenitor cells or circulating pro- genitors can give rise to vascular endothelial progenitor cells. Further studies are still needed to find more lung stem/ progenitor cell types in other locations of the pulmonary vasculature.

Like lung epithelial cells, a site-specific cell niche is required for the formation of different types of the pulmonary endothelial cells (Clark et al. 2008). In addition, some resident endothelial progenitors that are still unknown may form a universal but unspecified progenitor cell pool (Blaisdell et al. 2009).

The correct organization of the vascular plexus is important for both proper air- way branching morphogenesis and correct tissue perfusion in the lung (Warburton et al. 2010). Therefore, the cross talks between different lung cell compartments, mesothelium, mesenchyme, epithelium, and endothelium, can match the functions of both vascular and epithelial progenitor cells and are critical for lung development and repair/regeneration (Warburton et al. 2010). However, further studies are still required to identify the functional and biochemical properties and phenotypes of both endothelial cells and smooth muscle cells in the pulmonary vasculature.

2.3.2   Lung Smooth Muscle Stem and Progenitor Cells

The FGF10-expressing peripheral mesenchymal cell population serves as progeni- tor cells for the smooth muscle of the peripheral airway as shown in lineage-tracing studies (Mailleux et al. 2005; De Langhe et al. 2006; Ramasamy et al. 2007). The transdifferentiation of these progenitor cells to alpha smooth muscle actin fibers is regulated by both SHH and BMP4 that are expressed in the distal airways, indicat- ing that the sizing of peripheral airway smooth muscle progenitor cell population occurs early during lung formation (Mailleux et al. 2005; De Langhe et al. 2006; Kim and Vu 2006; Ramasamy et al. 2007). Meanwhile, another progenitor cell pop- ulation of airway smooth muscles was detected in the proximal mesenchymal tis- sues (Shan et al. 2008). Wnt2 is another important signaling pathway for the activation of the smooth muscle program by regulating the expression of myocar- din/Mrtf-B and Fgf10 in the lung (Goss et al. 2011).

Furthermore, progenitor cell populations for parabronchial smooth muscle cells exist at the distal branching tips (Mailleux et al. 2005). The differentiation of these progenitor cells to form smooth muscles is tightly regulated by highly interactive activating and inhibitory signal networks (Hogan 1999).

2.3.3   The Repair and Regeneration Capacities

Currently, there is a rudimentary understanding of the repair and regenerative capacities of lung mesenchymal stem/ progenitor cells in general. However, evidences suggest that the *chronic obstructive pulmonary disease* (COPD) is associated with a systemic process involving both trafficking of vascular endothelial progenitor cells (VEPCs) that engraft to the lung and activation of the bone marrow stimulating increased turnover (Lama et al., 2006; Lama and Phan, 2006). Once within the pulmonary tissue, VEPCs can differentiate into endothelial cells that contribute to tissue repair and maintenance and are governed by locally elaborated growth factors (Lama et al., 2006; Lama and Phan, 2006).

Chapter 3
Signals and Molecular Mechanisms Controlling Lung Stem/Progenitor Cell Development and Behavior

Abstract Better characterization and understanding of lung-specific stem cell behavior could lead to the discovery of new restoration solutions for normal and proper lung morphogenesis, repair, and regeneration. Recent data have accumulated on the behavior of these stem and progenitor cells such as self-renewal, fate, apoptosis, and differentiation into various cell types. Furthermore, many recent studies have focused on the modes of lung stem/progenitor cell division and the regulatory mechanisms of different aspects of lung stem/progenitor cell behavior, growth, and development. In this chapter, we describe recent advances on the factors, signals, and molecular mechanisms that control the self-renewing/proliferation, growth, and fate as well as differentiation of lung stem and progenitor cells.

Keywords Lung · Stem cells · Cell behavior · Cell fate · Wnt · Notch · TGF · Signaling pathway · Numb · Self-renewal · Differentiation

3.1 Signals and Genes Regulating General Lung Cell Growth and Development

The orchestration of lung development is by a tightly controlled interaction and cross talk between mesenchymal and epithelial cells. Many signals from mesenchymal cells are essential for branching morphogenesis of the epithelium in the lung (Warburton et al. 2010; Morrisey and Hogan 2010; El-Hashash 2013; Ibrahim and El-Hashash 2015; Schittny 2017; Spurlin III and Nelson 2017). Basically, there are five critical signaling pathways that control many biological processes during the embryonic development. These signaling pathways are Notch, Wnt, Hedgehog, transforming growth factor-b family (TGF-b), and fibroblast growth factor family (FGF).

The proper control of alveolar epithelial cells and other lung cell types is essential for normal development, morphogenesis, and function of the lung since insufficient capacity of the alveolar gas exchange is a major contributor to several serious lung diseases. The development and growth of different lung cells are tightly controlled by many intrinsic and extrinsic factors and signals, including transcription

© Springer International Publishing AG, part of Springer Nature 2018
A. El-Hashash, *Lung Stem Cell Behavior*,
https://doi.org/10.1007/978-3-319-95279-6_3

factors, growth factors, and other signaling molecules (Warburton et al. 2010; Morrisey and Hogan 2010; Schittny 2017; Spurlin III and Nelson 2017; El-Hashash 2018).

Transcription factors have a prominent role in lung formation and development as well as in lung repair and regeneration after injury (Warburton et al. 2010; El-Hashash 2013; Ibrahim and El-Hashash 2015; Berika et al. 2016; Schittny 2017; Spurlin III and Nelson 2017). Eya1 and Six1, for example, are important transcription factors for lung epithelial and mesenchymal cell development. They are expressed in distal lung epithelial and mesenchymal cells and act to optimize the activity of SHH signaling that negatively regulates FGF10 activity, to certain levels, which are sufficient for proper lung formation and morphogenesis (Ibrahim and El-Hashash 2015; Elshahawy et al. 2016). In addition, the activity and level of expression of SHH remarkably increase in the distal epithelial cells of *Eya1* or *Six1* knockout lungs, leading to a significant inhibition of the activity of FGF10 signaling, and consequently a server lung hypoplasia (Ibrahim and El-Hashash 2015; Elshahawy et al. 2016). Moreover, the ectopic expression of SHH leads to a clear reduction in the activity of FGF10 signaling pathway that results in abnormalities in the development of both lung mesenchyme and epithelium, as well as a defected branching morphogenesis (Bellusci et al. 1997). Similarly, the T-box transcription factor 2 (Tbx2) is a key regulator of the lung mesenchymal cell development. Tbx2 deficiency can result in sever hypoplastic lung mesenchymal cells that are characterized with a reduced cell proliferation, which results in a prematurely induced mesenchymal differentiation into fibrocytes (Lüdtke et al. 2013).

Growth factors are key regulators of lung epithelial cell development, and they may protect lung damages incurred upon alveolar cells (Ramasamy et al. 2007; Warburton et al. 2010; Ibrahim and El-Hashash 2015; Berika et al. 2016; Schittny 2017; Spurlin III and Nelson 2017; El-Hashash 2018). For instance, the mesothelium-derived FGF9 can regulate FGF10 signaling activities from the peripheral mesenchyme to the surrounding epithelial cells by the assembly of complex signaling elements that include FGFR2b, Ras, SHP2, Sos, and Grb2 in epithelial cells and Sprouty 2 (SPRY2) protein (Bellusci et al. 1997; Tefft et al. 2002, 2005; del Moral et al. 2006). In addition, both the mesenchymal-derived FGF9 and mesothelial-derived WNT2A signaling molecules are required for the maintenance of FGF-WNT/β-catenin signaling in the mesenchyme, while epithelial FGF9 mainly regulates the branching morphogenesis (Yin et al. 2011). Moreover, FGF9 signaling activities regulate the proliferation of mesenchymal cells, whereas β-catenin signaling activities are required for mesenchymal active FGF signaling in the mesenchyme (Yin et al. 2011).

Growth factors and other molecules are also essential for the development of lung mesenchymal cell derivatives, including smooth muscle and vascular endothelial cells as well as myofibroblast cells. The vascular endothelial growth factor (VEGF), for example, regulates the endothelial maintenance and morphogenesis, while the canonical BMP signaling shows high activity within the airway smooth muscle layer and the vascular network of the developing lung during the pseudoglandular stage (Tuder and Yun 2008; Sountoulidis et al. 2012). In addition, VEGF, erythropoietin and nitric oxide play a role in the processes of endothelial progenitor cell (EPC) mobilization and homing in the lung. Several related developmental changes occur after exposing newborn mice to hyperoxia, including the expression of VEGF, and erythropoietin receptor as well as endothelial nitric oxide synthase (Balasubramaniam et al. 2007). These changes also include a reduction of EPC number in the lung, bone marrow, and blood (Balasubramaniam et al. 2007). Furthermore, glycogen synthase kinase-3β/β-catenin signaling is another important regulator of the process of mesenchymal stromal cell differentiation into myofibroblast cells in the lung (Popova et al. 2012). However, more research is still required to identify molecular and cellular mechanisms as well as signaling pathways regulating mesenchymal cell differentiation in the lung.

3.2 Signals and Genes Regulating Lung Stem/Progenitor Cell Behavior: The Self-Renewing and Differentiation

The proper self-renewing and differentiation of specific lung stem and progenitor cells are critical for the maintenance of lung homeostasis. However, a remarkable feature of the airway stem and progenitor cells is their slow self-renewal and capacity to produce differentiated progeny in the lung. In addition, the behavior of stem and progenitor cells in the lung is a tightly controlled process and is regulated by both intrinsic signals and extrinsic factors, including transcription factors, growth factors, and a network of other signaling molecules.

3.2.1 Role of Transcription Factors

Like other systems, the self-renewing and proliferation of specific stem and progenitor cells in the lung are tightly controlled by many transcription factors and signaling molecules during the lung formation and morphogenesis. Mutations in these transcription factors or signaling molecules can lead to several severe human disorders, degenerative diseases, or cancer (Balasooriya et al. 2017). For example, the expression of thyroid transcription factor 1 (Ttf-1/Nkx2.1) transcription factor can mark the commitment of lung lineages and regulate the development of distal lung progenitor cells during early embryogenesis (Kimura et al. 1999). *Ttf1-/-* knockout mice have abnormal lungs with insufficient number of differentiation cells for survival (Kimura et al. 1996).

During lung embryogenesis, a stem and progenitor cell population at the distal epithelium can give rise to a bronchiolar-fated progeny and subsequently to an alveolar-fated progeny. One member of the STAT family of transcription factors, the signal transducer and activator of transcription 3 (STAT3), controls both the fate of lung alveolar cells and the bronchiolar-to-alveolar developmental transition

(Laresgoiti et al. 2016). In addition, the identity of lung epithelial cells is extrinsically determined during lung embryogenesis (Laresgoiti et al. 2016). Notably STAT3 transcription factor and glucocorticoid signaling can regulate both the timing of the early formation of alveolar cells and alveolar cell differentiation (Laresgoiti et al. 2016). Unexpectedly, neither STAT3 nor glucocorticoid pathway is absolutely required for specifying the fate of alveolar cells (Laresgoiti et al. 2016).

Sox9 is HMG box transcription factor that is highly expressed from E11.5 to E16.5 embryonic stages in the distal epithelial stem/progenitor cells (Liu and Hogan 2002). Sox9-specific conditional deletion in the lung, however, does not cause apparent defects in stem/progenitor cell behavior (Perl et al. 2005). This phenotype is probably because Sox9 probably functions redundantly with still unidentified factors that regulate stem and progenitor cell proliferation in the lung. Another transcription factor, N-myc, functions to maintain epithelial stem and progenitor cells in the distal lung (Okubo et al. 2005). N-myc may also promote the self-renewing of epithelial stem and progenitor cells (Okubo et al. 2005), while the mesenchymal nuclear factor 1 B (NF1B) controls both the proliferation and differentiation of epithelial cells during the lung maturity (Hsu et al. 2011).

The forkhead/winged helix (Fox) family members of transcription factors regulate the proliferation of stem/progenitor cells in the lung epithelium and show lung mutant phenotypes. Absence of both Foxa1 and Foxa2 transcription factors specifically in the lung can lead to a small formed lung that is characterized by a reduced cell division rate (Wan et al. 2005). In addition, Foxp1 and Foxp2 transcription factors are expressed at high levels in stem and progenitor cells of the distal lung epithelium and have a lung phenotype that is similar to Foxa1 and Foxa2 mutant mice. Moreover, Foxp22/2; Foxp11/2 transcription factors are key regulators of the maintenance of lung stem/progenitor cells and control their self-renewal divisions (Shu et al. 2007). Foxp22/2; Foxp11/2 double-mutant mice have hypoplastic lungs that are characterized by an inhibited cell proliferation (Shu et al. 2007).

The mesenchymal nuclear factor 1 B (NF1B) regulates the self-renewal and differentiation of airway epithelial cells during lung maturation (Hsu et al. 2011). In contrast, the transcription factor grainyhead-like 2 (GRHL2) can inhibit lung stem and progenitor cell differentiation and function as an organizer of mucociliary epithelium during the development of bronchial epithelial cells (Gao et al. 2013). Other transcription factors such as Six1 and Eya1 function to maintain epithelial stem and progenitor cells in the lung by controlling their proliferation and/or mode of division as well as the regulation of SHH-FGF10 signaling pathways (Berika et al. 2014; Ibrahim and El-Hashash 2015; Elshahawy et al. 2016).

3.2.2 Role of Growth Factors

Growth factors such as fibroblast growth factor family (FGF) and transforming growth factor-b family (TGF-b) are among the five critical signaling pathways that also include Notch, Wnt, and Hedgehog and control many biological processes during embryogenesis. In addition, they can drive differentiation and control the cell behavior of many types of stem cell lines. The functions of these signaling pathways in driving specific differentiation into various types of cells that have different functional roles in the lung have been recently intensively studied. These signaling pathways tightly interact and collaborate to regulate the proliferation of stem and progenitor cells at distal lung epithelium. However, the specific details of their interaction still need more studies.

Almost 22 FGFs and 4 FGF receptor (FGFR 1–4) families were reported in mammalian tissues. FGF signaling pathway plays essential roles in the lung lineage specification that occurs distal to the trachea (Serls et al. 2005; Ramasamy et al. 2007; De Langhe et al. 2008; El Agha et al. 2014). FGF10 acts as a chemotactic factor during epithelial branching in the lung. Lung mesenchymal cells express FGF10 that functions to maintain the proliferation of stem and progenitor cells in the lung epithelium. FGF10 overexpression can lead to the goblet cell metaplastic differentiation (Nyeng et al. 2008). As an upstream regulator of FGF signaling, retinoic acid signaling controls the primary lung bud formation and the expansion of stem and progenitor cells, by regulating Fgf10 expression levels through TGF-beta signaling in the lung (Chen et al. 2007).

FGFR2 receptor signaling acts to inhibit SOX2 expression to maintain epithelial stem/progenitor cells undifferentiated during lung development (Que et al. 2007; Abler et al. 2009; Volckaert et al. 2013). Surprisingly, the ectopic expression of Fgf10 in SOX2+ airway progenitors results in promoting the differentiation of basal cells that are stem cell populations in the developing lung (Rock et al. 2009; Mori et al. 2015; Watson et al. 2015). In addition, *Fgf10 signaling can* promote the differentiation of Sox2-expressing cells into the basal cell lineage (Volckaert et al. 2013). The overexpression of *Fgf10* can also induce Sox2-expressing proximal airway epithelial cell differentiation to form basal cells (Volckaert et al. 2013). This finding has been further tested in murine trachea by deleting one copy of FGFR2 receptor in the basal cells of the adult mouse that leads to both self-renewing and differentiation phenotypes, indicating the importance of FGFR2 signaling in both the self-renewing and differentiation of basal cells (Balasooriya et al. 2017). In addition, FGFR2 signaling may be important for maintaining Sox2 expression that is critical for the self-renewing and differentiation of basal cells (Balasooriya et al. 2017).

The RNA-binding protein (RBP) HuR regulates both FGF signaling activity and mesenchymal responses during lung branching morphogenesis (Sgantzis et al. 2011). HuR controls the expression levels of the *Fgf10* and *Tbx4* genes, and its deletion can abolish their regulation posttranscriptionally by the mesenchymal-derived FGF9 (Sgantzis et al. 2011). This can block the distal bronchial branching morphogenesis at the initiation of the pseudo-glandular stage (Sgantzis et al. 2011).

The BMP signaling pathway is one of the major regulators of lung epithelial stem and progenitor cell proliferation. Acting through BMP receptor 1 (BMPR1), BMP signaling pathway regulates cell division and survival as well as morphogenesis of the murine distal lung epithelial cells (Eblaghie et al. 2006). Deletion of either BMPR1A or its ligand, BMP4, in the lung epithelium, can lead to a hypoplastic lung formation, with reduced cell proliferation and a low expression level of the transcription factors: FoxA2 and N-myc (Eblaghie et al. 2006).

Some growth factors are major regulator of both the proliferation of mesenchymal cells and smooth muscle cell (SMC) differentiation. For example, inhibiting the activity of *Fgf9* can suppress cell proliferation in the lung mesenchyme, leading to a reduction of epithelial branching morphogenesis in the lung (Colvin et al. 2001). In addition, treating explants of the embryonic lung mesenchyme with FGF9 peptides in culture can inhibit SHH-controlled mesenchymal cell differentiation into SMCs, without affecting the proliferation of mesenchymal cells (Weaver et al. 1999). Furthermore, FGF9 signaling activities in the lung mesenchyme are, therefore critical for the maintenance of mesenchymal progenitors in an undifferentiated state. Similarly, FGF10 signaling pathway coordinates vascular development and the formation of alveolar smooth muscle cells (Ramasamy et al. 2007).

3.2.3   Role of Micro-RNAs (miRNAs)

miRNAs are noncoding and small RNAs (~22 nucleotides long) that control the expression of genes posttranscriptionally. In addition, miRNAs are key inhibitors of the translation of target mRNAs and present as potential disease mechanisms, biomarkers, and therapeutic targets. Early miRNAs' studies have led to the discovery of the inhibitor of the LIN-14 protein that is called *Lin-4* in *Caenorhabditis elegans* and then the discovery of let-7. *Lin-4* and let-7 miRNAs control the timing of development in *C. elegans* (Lee et al. 2014; Reinhart et al. 2000). More than 1,000 validated miRNA genes have been discovered and characterized in the human genome, in which they regulate different cell biology processes, including cell proliferation, differentiation, migration, and death during mammalian development (Sayed and Abdellatif 2011 and Friedle human genomedate). Interestingly, one miRNA can regulate hundreds of downstream targets, while multiple miRNAs are able to function cooperatively to silence the same target gene (Lu and Clark 2012). In addition, the abnormal pattern of miRNA expression levels may result in an abnormality in the expression levels of their downstream target genes, which may eventually lead to the development of human disease. While certain miRNAs can control several epigenetic regulators, the epigenetic phenomena are major factors that may cause miRNA dysregulation (Sato et al. 2011a, b; Ameis et al. 2017).

The dysregulations of miRNAs may occur in many human diseases, including lung cancer, and, therefore, they may play key roles in pediatric respiratory disease, idiopathic lung fibrosis, and cardiovascular diseases (Croce 2009; Quiat and Olson 2013; Pandit and Milosevic 2015), as well as both psychiatric diseases and many childhood diseases (Miller and Wahlestedt 2010; Omran et al. 2013). Further studies are needed to understand both the biological and functional roles of miRNAs in different types of pediatric respiratory diseases, including asthma, bronchopulmonary dysplasia and cystic fibrosis.

miRNAs are important regulators of the behavior of epithelial stem/progenitor cells and lung development (Lu et al. 2007; Carraro et al. 2009). For example, the

overexpression of the entire miR-17–92 cluster throughout the developing lung epithelium can lead to an increase in the number of highly proliferative multipotent
epithelial progenitor cells and a delayed epithelial differentiation (Lu et al. 2007).
miR-17–92 may, therefore, promote self-renewal of stem/progenitor cells at the
expense of differentiated cell divisions (Lu et al. 2007).

Some miRNAs such as miR-17 and its paralogs, miR-20a and miR-106b, are
detected in the developing lung at the pseudo-glandular phase, target the activities
of both the signal transducer and activator of transcription 3 (Stat3) and mitogen-
activated protein kinase 14 (Mapk14), and thus have a major role in the lung formation and development (Carraro et al. 2009). Downregulation of these three miRNAs
simultaneously in isolated lung epithelial explants grown in culture can alter
FGF10-induced budding morphogenesis (Carraro et al. 2009). Inhibiting these three
miRNAs can also lead to other biological changes, including the reduction of
E-cadherin expression levels, changes in E-cadherin distribution pattern, augmenta-
tion of beta-catenin activity, and enhanced expression levels of both Fgfr2b and
Bmp4 (Carraro et al. 2009). Interestingly, the overexpression of Stat3 and Mapk14
simultaneously in isolated lung epithelial explants can mimic the distributional
changes of E-cadherin that occur after the downregulation *of miR-17, miR-20a, and
miR-106b*. This suggests that miR-17 family acts to modulate FGF10-FGFR2b
downstream signaling elements through targeting the activity of Stat3 and Mapk14,
hence controlling the expression level and distribution of E-cadherin that affect the
morphogenesis of lung epithelial buds in response to FGF10 signaling activity during lung development (Carraro et al. 2009).

Other types of micro-RNAs are also important for the behavior of lung stem and
progenitor cells during normal development and cancer formation (Tian et al. 2011;
Fan et al. 2016*).* The *miR302/367* cluster of micro-RNA regulates the balance
between cell proliferation and differentiation of early developing murine endodermal progenitor cells, under the control of Gata6 transcription factor in the lung
(Tian et al. 2011). In addition, miR-17–92 is abundant in undifferentiated progenitor
epithelial cells and induces their proliferation by targeting the retinoblastoma-like 2
(Rbl2) cell cycle regulator (Lu et al. 2007, 2008). Furthermore, miR-127 and miR-
351 may regulate mesenchymal to epithelial transition (EMT) during lung development (Bhaskaran et al. 2009). The activities of lung micro-RNAs are controlled by
several factors and mechanisms, including the epithelial sodium channels (ENaC;
Ding et al. 2017).

3.2.4 Role of Other Factors and Signal Molecules

A network of signaling molecules between the cells regulates the embryonic development and controls the behavior of tissue-specific stem and progenitor cells. These
signaling molecules belong mainly to five major signaling pathways that are Wnt,
Notch, and Hedgehog, besides FGF and TGF-beta growth factors.

Both Wnt2/2b and beta-catenin signaling are essential for the specification of lung stem and progenitor cells in the foregut (Goss et al. 2009). Wnt2/2b knockout mouse embryos show complete lung agenesis and undetectable expression levels of the earliest lung endodermal marker Nkx2.1 (Goss et al. 2009). This phenotype is also shown upon the deletion of beta-catenin specifically in the endoderm, suggesting that Wnt2/2b canonical signaling pathway functions to regulate lung endodermal stem/progenitor cell specification within the developing foregut (Harris-Johnson et al. 2009; Goss et al. 2009). In addition, Wnt2 signaling is required for the activation of the program of airway smooth muscle by regulating the expression levels of both Fgf10 and myocardin/Mrtf-B (Goss et al. 2011). Furthermore, the expression of Wnt ligand, Wnt5a, is detected in the distal epithelial tips of the embryonic lung, in which regulates the epithelial cell proliferation. Wnt5a/knockout mouse lungs show an increased epithelial cell proliferation, leading to the formation of additional branches of the conducting airway epithelium (Li et al. 2002). Whether defects in the behavior of lung stem and progenitor cells can lead to this Wnt5a-/- lung phenotype still needs more research.

The Wnt signaling pathway controls the behavior of alveolar epithelial type II progenitors and, therefore, plays important functions in the tight regulation of pulmonary alveologenesis that is essential for proper lung development, morphogenesis, and function (Warburton et al. 2000, 2010). The Wnt signaling regulates the expansion of the alveolar epithelial type II cell population through promoting cell proliferation and controlling the balance between type I and type II alveolar epithelial cells (Frank et al. 2016). Using the Wnt signaling reporter system, the axin2+ alveolar type 2 cells ($AT2^{Axin2}$), it was found that the number of $AT2^{Axin2}$ cells increases in correlation with increased Wnt signaling activities during lung alveologenesis (Frank et al. 2016). $AT2s^{Axin2}$ also induces the growth of alveolar epithelial type II cells during the alveolus formation (Frank et al. 2016). Enhanced Wnt signaling activity can, therefore, lead to the expansion of alveolar epithelial type II cell populations, whereas inhibited Wnt signaling can direct the development of the alveolar epithelium toward the formation alveolar epithelial type I cell lineage (Frank et al. 2016).

Beta-catenin signaling is also a major regulator of the behavior of mesenchymal stem and progenitor cells in the lung. Beta-catenin signaling controls mesenchymal-derived FGF signaling pathway, which in turn, regulates the proliferation of mesenchymal cells in the developing lung (Yin et al. 2011). Similarly, the production of myofibroblasts by differentiation of mesenchymal stromal cells is regulated by the glycogen synthase kinase-3beta/beta-catenin signaling in the neonatal murine lung (Popova et al. 2012).

Both Notch and SHH signaling are important for the development of lung-specific stem/progenitor cells. Canonical Notch signaling has a major regulatory function in the lung cell fate and, therefore, controls the selection of ciliated versus Clara cell fate in the developing lung (Morimoto et al. 2010). SHH is expressed by the lung distal epithelial cells and regulates both cell proliferation and branching morphogenesis (Pepicelli et al. 1998; Kugler et al. 2015). In addition, ectopic overexpression of Gli2 that mediates SHH activity in lung mesenchymal cells can lead to increased activities of SHH as well as enhanced cell proliferation by regulating cyclin expression (Rutter et al. 2010). In addition, *SHH signaling can* inhibit the expression of FGF*10* in the growing epithelial buds in the developing lung (Pepicelli et al. 1998; Chuang et al. 2003).

Other growth factors are also important for both normal and pathological development of lung-specific stem and progenitor cells. For example, the vascular endothelial growth factor (VEGF) and nitric oxide as well as erythropoietin are important for cell mobility and homing of lung endothelial progenitors (EPCs) during the development of the bronchopulmonary dysplasia (BPD) and other developmental disorders in the lung (Stevens et al. 2008). Moreover, studies on the functional role of endothelial progenitor cells in the BPD show that oxygen toxicity can disrupt the cell growth of lung vascular and alveolar compartments, which limits the surface area of the gas-exchange region in the lung (Balasubramaniam et al. 2007). Other associated changes include a reduced expression level and activity of VEGF and erythropoietin receptors as well as a decrease in both the endothelial nitric oxide synthase and EPC number in the blood and bone marrow (Balasubramaniam et al. 2007).

3.2.5   Role of the Epigenetic Factors

The epigenetic factors are important for the proper commitment of the *alveolar type I epithelial cell* lineage, which is primitive units of the alveolar epithelium essential for both normal lung formation and the postnatal breathing. More recent studies have focused on how stem and progenitor cells in the distal lung epithelium differentiates and produce the alveolar epithelial type I cells prior to the birth and how the developmental niche instructed by neighboring mesenchymal cells affects this developmental process (Wang et al. 2016). In addition, early studies demonstrated that the epigenetic factors, histone deacetylases (HDAC), act to modulate both the gene expression and chromatin structure by the deacetylation of histones and non-histone proteins (Wang et al. 2013). In addition, epithelial histone deacetylases 1/2 play an important role in the development of Sox2+ proximal lung endodermal progenitor cells during lung formation and in the airway secretory cell regeneration during postnatal stages of lung development (Wang et al. 2013).

The lung mesenchyme expresses the histone deacetylases 3 that regulates the spreading of alveolar epithelial type I cells through a micro-RNA-TGFβ signaling axis during lung sacculation (Wang et al. 2013). Specific deletion of the histone deacetylase 3 (HDAC3) in the lung mesenchyme uncovers HDAC3 functions in the regulation of mesenchymal cell proliferation (Wang et al. 2016). HDAC3 also regulates the differentiation of alveolar epithelial cells during lung development (Wang et al. 2016). In addition, the HDAC3 deletion in the lung mesenchymal cells can lead to both inhibition of Wnt/β-catenin activities in the lung epithelium and defects in the alveolar epithelial type I cell differentiation, suggesting a regulatory role of the HDAC3 for Wnt signaling activities and cell differentiation in the lung epithelium (Wang et al. 2016).

The epigenetic factor Ezh2 is also important for lung stem/progenitor cell behavior. The Ezh2-dependent mechanism is critical for the developing lung mesothelium (Snitow et al. 2016). This Ezh2-dependent mechanism acts to restrict the smooth muscle gene expression program, and, therefore, it allows proper cell fate decisions to occur in the multipotent mesodermal lineages (Snitow et al. 2016). In addition, the PRC2 component Ezh2 is critical for restricting lung smooth muscle cell differentiation in the developing mesothelium of the mouse embryo (Snitow et al. 2016). Furthermore, the mesodermal loss of Ezh2 in the lung can lead to the formation of ectopic smooth muscle cells in the sub-mesothelial region of the developing mesoderm (Snitow et al. 2016). Interestingly, deletion of Ezh2 specifically in the lung mesothelium confirms the mesothelial cell-autonomous functional roles of Ezh2 in the suppression of the differentiation of smooth muscle cells during murine lung development. One potential mechanism for Ezh2 effects on smooth muscle cell differentiation is by suppressing the expression of myocardin and Tbx18 that control the differentiation of the mesothelium into smooth muscle cell types (Snitow et al. 2016). Interestingly, the epigenetic changes in the lineages of the mesoderm of the adult lung may contribute to the development of sever human disorders or diseases, including the idiopathic pulmonary fibrosis (IPF) and the chronic obstructive pulmonary disease (COPD; Snitow et al. 2016).

3.3   Regulatory Signal Mechanisms of Stem and Progenitor Cells During Lung Repair and Regeneration

Growth factors may protect lung damages incurred upon alveolar cells, besides their roles as major regulator of lung epithelial cell development (Ramasamy et al. 2007; Warburton et al. 2010; El-Hashash 2013; Ibrahim and El-Hashash 2015; Berika et al. 2016). FGF family of growth factors is important for both lung development and the protection of alveolar cells against lung injury, and the amelioration of DNA damages by FGF7 in the alveolar epithelium is a good example (Buckley et al. 1997). In addition, FGF7 can enhance both the protection of mitochondria and the alveolar epithelial cell ability for migration and repair in a scratch assay in culture (Buckley et al. 1997). In animal models, treatment with FGF7 can enhance the cell resistance to in vivo alveolar injury (Ray et al. 2003; Plantier et al. 2007). Furthermore, FGF10 is another FGF family member that has a protective functional role in both lung fibrosis and injury (Gupte et al. 2009).

Transcription factors are also important for the repair and regeneration of the lung after injury by regulating the cell behavior. For example, E74-like transcription factor-3 (Elf3) regulates both cell proliferation and differentiation in the lung during the repairing process of the bronchiolar airway epithelial cells after a specific injury in the Clara cells (Oliver et al. 2011). Moreover, the NK2 homeobox 1 (NKX2-1) that is also called the thyroid *transcription factor 1 (TTF-1)* can modulate and possibly initiate the compensatory lung growth at its early stage (Takahashi et al. 2010).

Furthermore, the nucleoside, inosine, shows protective functions against oxygen lung injury such as mitochondrial protection, glutathione repletion, and enhanced expression of VEGF as well as reduced apoptosis (Buckley et al. 2005). Protecting or enhancing the functions of alveolar stem/progenitor cells is probably an effective therapeutic approach for lung injury that should be assessed in clinical trials in the future (Buckley et al. 2005).

In summary, many signaling pathways and factors interact to regulate the formation, behavior, function, and repair/regeneration of lung stem and progenitor cells. However, more studies are still needed to better understand how these signaling pathways interact and collaborate to do their functions in the lung.

Chapter 4
Signals and Molecular Mechanisms Regulating Stem Cell Behavior in Other Systems (e.g., Hematopoietic Stem Cells)

Abstract Over recent years, many *studies* of stem cell behavior in different organs *have accumulated* evidences on the molecular mechanisms and signaling pathways that regulate key aspects of stem/progenitor cell behavior such as self-renewal, differentiation, and apoptosis. These molecular mechanisms include Wnt, TGF beta, and Notch signaling pathways that are well investigated in hematopoietic stem cells. In this chapter, we will discuss the hematopoietic stem/progenitor cell niche and both the molecular mechanisms and signaling pathways that control their behavior since this will improve our understanding of the role of these mechanisms and signals in the behavior of lung stem/progenitor cells that is still not well studied compared to hematopoietic stem/progenitor cells.

Keywords Stem cells · Cell behavior · Cell fate · Hematopoietic stem cells · Wnt · Notch · TGF · Signaling pathway

Lifelong maintenance of organ functions requires a continual replacement of cells. This is accomplished through the resident stem cell populations and is illustrated particularly in tissues with high turnover rates such as the skin. In this chapter, we will focus on the functions of various types of signaling pathways and molecular mechanisms in the control of the behavior of hematopoietic stem/progenitor cells, including cell self-renewal and differentiation, since they are well studied in these stem cell types.

Stem and progenitor cells that exist in the adult bone marrow could form *different blood cell* types (*lineages*) that have *various* functions. These hematopoietic stem and progenitor cells are capable of balancing their self-renewal with differentiation to maintain blood populations and stabilize the resident stem cell pool within the bone marrow. Like other stem cell types, the symmetric and asymmetric divisions are the two known modes of cell divisions for self-renewing hematopoietic stem and progenitor cells. This self-renewal capability is a critical aspect for lifelong hematopoiesis in humans and other organisms. The asymmetric mode of self-renewing division takes place when the resultant two daughter cells have two different fates, with one daughter acquires the stem cell fate and the other gains a differentiation fate. This asymmetric mode of cell division allows for a steady

© Springer International Publishing AG, part of Springer Nature 2018   27
A. El-Hashash, *Lung Stem Cell Behavior*,
https://doi.org/10.1007/978-3-319-95279-6_4

population of hematopoietic stem and progenitor cells and is particularly important in the healing of hematopoietic injury or after transplantation. Whereas, the symmetric mode of cell division occurs when stem cells undergo a self-renewal division that leads to the production of two daughter cells, both of which maintain the stem cell properties. Since both daughters become stem cells, this symmetric mode of cell division may also lead to the expansion of the pool of stem cells (Berika et al. 2016).

4.1   Hierarchy and Heterogeneity of Hematopoietic Stem and Progenitor Cells

Hematopoietic stem cells normally produce a progeny that eventually loses its self-renewing capacity and differentiates into mature cells. Notably, the bone marrow-derived hematopoietic stem/progenitor cells cannot characteristically express many surface markers that are normally expressed by mature blood cells. A good example is Sca1 and c-kit that are known as the LSK compartments (lineage-/Sca1+/c-kit+) and are highly expressed by hematopoietic stem and progenitor cells (Spangrude et al. 1988; Okada et al. 1992). In irradiated mice, all blood cell lineages are noted to be restored by several groups, when a single hematopoietic stem cell was transplanted into these mice (Osawa et al. 1996; Matsuzaki et al. 2004). Remarkably, 100 LSK cells are sufficient for repopulating blood cell lineages in lethally irradiated mice for a long period of time (Okada et al. 1992).

There are two types of reconstituting hematopoietic stem and progenitor cells: long-term and short-term cells. Short-term hematopoietic stem and progenitor cells show a limited capacity for self-renewal and can generate blood cells for a specific amount of time in vivo. In contrast, long-term hematopoietic stem and progenitor cells have an unlimited capacity for self-renewal and, therefore, can sustain lifelong hematopoiesis. The LSK compartment contains both types of reconstituting hematopoietic stem cells, short-term and long-term, and multipotent progenitors (Morrison and Weissman 1994; Adolfsson et al. 2001; Yang et al. 2005). Notably, several recent trials showed that the isolation of human hematopoietic stem and progenitor cells is difficult, probably due to the lack of a proper in vivo assay. For example, mouse strains that are nonobese diabetic/severe combined immune deficient (NOD/SCID) generally have low engraftment of human cells and, therefore, do not live long enough for a long-term analysis. To overcome these problems, some researchers have performed serial transplantations for assaying NOD/SCID mouse strains. However, the results and findings of these studies may be unreliable because hematopoietic stem/progenitor cells are kept under constant stress in humans.

4.2   Hematopoietic Stem and Progenitor Cell Niche

The hematopoietic stem/progenitor cells start their development in the fetal liver, before establishing their localization in the bone marrow. Interestingly, primitive or premature hematopoietic stem cells can cycle exponentially in the fetal liver and then transform from this state to a quiescence state in 3 to 4 weeks' time through an intrinsic mechanism in mice (Bowie et al. 2006).

The hematopoietic stem and progenitor cell niche is a complex system that is controlled and affected by many intrinsic and extrinsic factors, which are currently under intensive research investigations. There are two types of connective tissue that line the bone: the periosteum that lines the outer layer of bones while the endosteum lines the inner cavity or the bone marrow. The hematopoietic stem and progenitor cell niche has been identified on the endosteal surface of the bone. The hematopoietic stem/progenitor cell niche is composed of the microenvironment and cells that underlie the basement membrane, where hematopoietic stem/progenitor cells develop and multiply (Schofield 1978).

Niche cells that are underneath the basement membrane produce signals that can block the cell differentiation and regulate the cell division of hematopoietic stem cells. These stem cells can divide either symmetrically or asymmetrically, a process that is regulated by many local signals and factors (reviewed in Morrison and Scadden 2014). One of the key components of this niche is osteoblasts that form bone tissues. Some studies have provided evidences that the lack of osteoblasts in vivo can lead to a decrease of the hematopoietic cellularity, which support the conclusion that osteoblasts are critical for the microenvironment of hematopoietic stem cells (Wilson and Trumpp 2006). There are several other constituents that influence the niche of hematopoietic stem and progenitor cells such as soluble factors, matrix proteins, and cellular ligands (Stier et al. 2005; Nilsson et al. 2005). The calcium gradient is another factor that is essential for maintaining the localization of hematopoietic stem and progenitor cells in their respective niche and region in the bone marrow (Adams et al. 2006). In addition, the environmental stress can remarkably affect the specific niche of hematopoietic stem/progenitor cells (reviewed by Morrison and Scadden 2014).

One of the characteristic features of the hematopoietic stem/progenitor cell niche is that it is constantly in a state of hypoxia (Parmar et al. 2007). It is well established that the oxidative stress can lead to an early bone marrow failure, and, therefore, the oxygen level must be kept low. If the oxygen level abnormally increases, it may result in an increase of the frequency of hematopoietic stem/progenitor cell cycling, thereby exhausting their functionality at a faster rate (Ito et al. 2004, 2006). Therefore, the maintenance of a hypoxic microenvironment is essential for the proper structure and function of the niche of hematopoietic stem and progenitor cells.

4.3   Molecular Control of Hematopoiesis

Recent studies have revealed more about the molecular mechanisms and signaling pathways that control the behavior of different stem cell types, including hematopoietic stem and progenitor cells. Several developmentally conserved signaling pathways were reported to play critical roles in the regulation of the behavior of stem cells, including the Notch, Smad, Sonic hedgehog (Shh), Wingless-type (Wnt), and BMB signaling pathways. Identification of the functions of these signaling pathways has enhanced our knowledge on the development and function of various types of stem cells and their role in human health and diseases.

Many intrinsic and extrinsic signals and factors can regulate hematopoiesis to accommodate for the acute blood loss or infection or for a variety of other cases and conditions (Till et al. 1964; Metcalf 1993; Enver et al. 1998). In the last two decades, the functions of many molecular mechanisms, signaling molecules, and factors, including growth factors, transcription factors, cytokines, cell-cycle regulators, and chromatin modifiers, have been determined in the control of the development, self-renewal, and fate of hematopoietic stem and progenitor cells (Berika et al. 2016). Discovering the functions of these mechanisms and signal is critical in the field of hematopoietic stem cells and will be summarized below.

4.3.1   Cytokines as Critical Regulators of Hematopoietic Stem/ Progenitor Cells

Accumulated studies have tried to use classic hematopoietic cytokines to expand hematopoietic stem and progenitor cells in culture. However, many of these studies have failed to prevent the hematopoietic stem and progenitor cell differentiation in culture. Interestingly, the repopulation of murine hematopoietic stem and progenitor cells in a 10-day culture protocol containing interleukin (IL)-11, Flt-3 ligand, and stem cell factor (SCF) was reported two decades ago (Miller and Eaves 1997). However, the culture conditions used in these early studies were not sufficient to expand hematopoietic stem and progenitor cells derived from the fetal liver (Nandurkar et al. 1997; Adolfsson et al. 2001; Christensen and Weissman 2001; Audet et al. 2001). These studies have demonstrated that hematopoietic stem and progenitor cells lack the expression of the FMS-like tyrosine (Flt-3) receptor and their self-renewal capacity may be regulated by the gp130 protein. The gp130 protein is a transmembrane protein that functions as a founding member of the class of all cytokine receptors and, therefore, is a part of the receptors for IL-6 and IL-11. In addition, these studies showed that the suppression of the IL-11 receptor does not apparently affect hematopoiesis (Nandurkar et al. 1997; Adolfsson et al. 2001; Christensen and Weissman 2001; Audet et al. 2001).

Among the cytokines examined using purified murine hematopoietic stem cells are *thrombopoietin* (*TPO*) and stem cell factor (*SCF*). TPO is an important factor that induces the development of platelet precursor cells (megakaryocytes). TPO stimulation of megakaryocyte development results in increased numbers of circulating platelet cells similar to the stimulatory effect of the erythropoietin (EPO) on erythroid precursor cells (Kaushansky and Drachman 2002). Many studies have found that the receptors for TPO and SCF are both expressed by the repopulating hematopoietic stem cells (Ikuta and Weissman 1992; Adolfsson et al. 2001; Yang et al. 2005; Kimura et al. 1998; Solar et al. 1998; Buza-Vidas et al. 2006). Mutations in the TPO gene or its receptor, c-mpl, lead to a remarkable reduction of hematopoietic stem and progenitor cell number in mice (Kimura et al. 1998; Solar et al. 1998). In addition, TPO functions to support the viability and suppresses apoptosis in hematopoietic stem and progenitor cells (Borge et al. 1996; Pestina et al. 2001). Later studies have suggested that TPO functions to counteract apoptosis rather than promoting the expansion of hematopoietic stem/progenitor cell compartment (Buza-Vidas et al. 2006).

The lymphocyte *adapter* protein (*LNK*) belongs to a family of *intracellular* adaptor proteins and functions to inhibit certain cytokine signaling elements, including IL-3, SCF, TPO, IL-7, and erythropoietin pathways (Takaki et al. 2000; Velazquez et al. 2002; Tong et al. 2005; Buza-Vidas et al. 2006; Seita et al. 2007). The *LNK* knockout mice are characterized by an increase of hematopoietic stem and progenitor cell number and an enhanced repopulation function (Ema et al. 2005; Buza-Vidas et al. 2006; Seita et al. 2007). These studies suggest that the function of *LNK* is to inhibit the TPO signaling, which is an important positive regulatory pathway of hematopoietic stem and progenitor cells.

4.3.2   Notch Signaling Pathway and Hematopoietic Stem and Progenitor Cells

The Notch is a well-known signaling pathway that plays important roles in lineage specifications and both the differentiation and self-renewal of stem and progenitor cells in different systems. The Notch signaling pathway is also a critical regulator of lymphocyte development and function (Radtke et al. 2004). The transcripts of both Notch and Notch ligand are expressed in primitive hematopoietic cells and other cells in the putative microenvironment of hematopoietic stem/progenitor cells (Karanu et al. 2000; Milner et al. 1994). The Notch ligands, Delta and Jagged, are capable of expanding human and murine hematopoietic progenitor cells in culture (Karanu et al. 2000, 2001; Ohishi et al. 2002; Varnum-Finney et al. 2003; Vas et al. 2004; Delaney et al. 2005; Suzuki et al. 2006). In addition, enforcement of the Notch signaling activity or its downstream target Hes-1 can lead to immortalization of primitive hematopoietic stem/progenitor cells (Varnum-Finney et al. 2000; Stier et al. 2002; Kunisato et al. 2003).

Genetically modified mice to produce activated PTH/PTHrP receptors (PPRs) that are specific for osteoblast cells were assessed in a remarkable study by Calvi et al. (2003). They found that PPR stimulation increases both the number of osteo-blasts and their production of the Notch ligand jagged 1. PPR stimulation also enhances both hematopoietic stem cell number and Notch1 signaling activity in vivo (Calvi et al. 2003). In addition, activation of PPR by its specific ligand, parathyroid hormone (PTH), can enhance osteoblast cell number in stromal cell cultures and increase the ex vivo growth of primitive hematopoietic cells that is revoked by the blocking of Notch activity with γ-secretase inhibitors (Calvi et al. 2003). Furthermore, injection of parathyroid hormone (PTH) into wild-type animal models can enhance the number of stem cells (Calvi et al. 2003). Osteoblastic cells in the niche of hematopoietic stem cells are, therefore, major regulators of the functions of these stem cells through the activation of Notch signaling pathway in vivo (Calvi et al. 2003). Notably, hematopoietic stem and progenitor cells that are deficient in Notch-1 could reconstitute Jagged-1 deficient hosts, possibly through a compensa-tory mechanism with other Notch receptors and ligands. The elimination of all Notch signaling pathways would, therefore, help in figuring out the hematopoietic stem cell expansion ex vivo in the future (Mancini et al. 2005).

4.3.3   The Wnt Signaling as a Key Regulatory Pathway of Hematopoietic Stem and Progenitor Cell Self-Renewal

The Wnt proteins have been detected in both fetal and adult bone marrows. Studies that utilized Wnt proteins in controlled reconstitution assays have shown an increase in the self-renewal process of hematopoietic stem/progenitor cells grown in culture (Willert et al. 2003). However, the exact mechanisms of operation of the Wnt path-way in hematopoietic stem and progenitor cells still need more clarifications. The transcription of target genes that are influenced by the expression of β-catenin rep-resents the product of Wnt signaling. Surprisingly, the expression of β-catenin in vivo exhibits a phenotype that contradicts its counterpart in culture (Reya et al. 2003; Cobas et al. 2004; Kirstetter et al. 2006). This difference supports several previous studies that have demonstrated the complexity of the in vivo environment and suggested backup or alternative molecular mechanisms that counter the impair-ment of β-catenin in the signaling pathway (Reya et al. 2003).

Both the Wnt and Notch signaling are among a wide range of many other path-ways that regulate the development of hematopoietic stem and progenitor cells, and the interaction between these two signaling pathways in hematopoietic stem cells is well studied. The Wnt signaling pathway apparently promotes the proliferation of the LSK compartments, while the Notch pathway ensures that these cells remain undifferentiated (Duncan et al. 2005). Thus, it is notable that the Wnt and Notch signaling pathways are co-dependent as steps in each pathway tend to overlap, in which respective ligands influence and interact with ligands belonging to the other pathway (Trowbridge et al. 2006).

4.3.4 The Smad, TGF-β, and BMP Signaling Pathways and Hematopoietic Stem and Progenitor Cells

Little is known about the functions of the Smad signaling pathways in hematopoietic stem/progenitor cells and other cell types, mostly because the Smad pathway consists of some overlapped signal transduction pathways that may diverge or converge. Therefore, the Smad pathway has yet to be constructed and fully understood in various tissues and systems. The survival of the Smad knockout mice is another problem that renders the study of the Smad pathway difficult in different systems. Despite these difficulties, some studies have demonstrated that the Smad signaling pathway negatively regulates the growth of hematopoietic stem and progenitor cells in vivo (reviewed in Blank and Karlsson 2011).

Both the transforming growth factor- β (TGF-β) and bone morphogenetic proteins (BMPs) play major roles in mediating both the proliferation and self-renewal capabilities of hematopoietic stem and progenitor cells. TGF-β treatment can markedly reduce the hematopoietic stem and progenitor cell growth, while BMPs treatments effectively maintain their self-renewal rate in culture (Sitnicka et al. 1996; Garbe et al. 1997; Batard et al. 2000). Conversely, in vivo studies have shown contradictory results with many of these studies lack the expected in vitro phenotype. These inconsistencies suggest that alternative molecular mechanisms exist and generate redundancy in the Smad signaling pathway (Larsson et al. 2003, 2005). The redundancy of the Smad signaling in hematopoietic stem/progenitor cells has been supported by other in vivo studies that identified and characterized the functions of additional ligands that arbitrate signals in the Smad pathway, including Smad 7 and Smad 4 that demonstrate opposing regulatory functions. Thus, while Smad 7 positively controls the self-renewing process, Smad 4 shows a negative regulatory effect on the populations of hematopoietic stem and progenitor cells (Nishita et al. 2000; Labbé et al. 2000; Itoh et al. 2004). Yet, the exact mechanisms by which these ligands can interact with each other or with other intermediaries have not been clearly identified. In addition, more research is still needed to assess how different combinations of ligands influence hematopoietic stem/progenitor cells and how they regulate their self-renewal, development, and behavior.

4.3.5 The Angiopoietin-Like Protein Functions in Hematopoietic Stem and Progenitor Cells

The growth of hematopoietic stem and progenitor cells occurs substantially in the fetal liver during its early stages of development. The angiopoietin-like (Angptl) growth factor proteins promote the amplification of hematopoietic stem and progenitor cells in the fetal liver. However, there is currently a rudimentary knowledge of the physiological roles of these Angptl proteins, and the molecular mechanisms by which they expand hematopoietic stem and progenitor cells still need further investigations. However, the available data from recent studies provide evidences that there is a reduction in the number and quiescence of hematopoietic stem and progenitor cells in Angptl3 deficient mice. In addition, there is a defect in the repopulation ability of hematopoietic stem cells that are transplanted into Angptl3-null recipient mice (reviewed by Zheng et al. 2011). Moreover, Angptl3 is intensively expressed by the bone marrow sinusoidal endothelial cells that are localized close to hematopoietic stem and progenitor cells. Notably, Angptl3 deficiency in either bone marrow stromal cells or endothelial cells can cause a severally reduced capability of these cells for expanding the repopulating hematopoietic stem/progenitor cells. Mechanistically, the Angptl3 functions to inhibit the expression Ikaros, a transcription factor that controls the repopulation of hematopoietic stem cells. As an extrinsic factor, the Angptl3 can, therefore, regulate the stemness of hematopoietic stem cells in the bone marrow (reviewed in Zheng et al. 2011).

Chapter 5
Mode of Cell Division as a Regulatory Mechanism for Lung Stem Cell Behavior Compared to Other Systems

Abstract Over the last decade, new insights have been accumulated on the cell fate, behavior, and mode of division of stem/progenitor cells in the respiratory system and many other systems. The balance between symmetric and asymmetric divisions of tissue-specific stem cells is essential for proper organ formation, development, function, and repair/regeneration and, therefore, is tightly controlled. Improper asymmetric cell division can disrupt the tightly regulated morphogenesis of organs, while uncontrolled symmetrical cell division can promote tumor formation. In this chapter, we describe recently accumulated data on the division mode (symmetric vs. asymmetric) of lung stem/progenitor cells. In addition, we describe and compare these division modes and their regulatory mechanisms between the lung and other tissue-specific stem cells in mammals and other organisms, in which these processes are well investigated.

Keywords Lung · Stem cell · Cell behavior · Symmetric · Asymmetric · Cell division · Muscle stem cells · Epidermal stem cells · Neural stem cells

The ability to dictate cell fate decisions is crucial for the normal development of different tissues and organs in animals and humans. In addition, the continuous execution of this crucial process ensures proper tissue and organ development, repair, regeneration, and homeostasis throughout life, whereas abnormalities in the molecular machinery that control this process may contribute to diseases.

Multipotent epithelial stem cells are localized at the embryonic distal epithelial buds and developing airways during lung embryogenesis and development (Rawlins and Hogan 2006; Rawlins 2008; Rawlins et al. 2009a, b). The control of the behavior of epithelial stem and progenitor cells is essential for proper lung formation, development, and function, like other organs (Warburton et al. 2000, 2010; El-Hashash 2013; Ibrahim and El-Hashash 2015). A significant stem/progenitor cell deficiency can, therefore, explain several lung defects or disorders such as common lethal disorders/defects of the capacity of gas diffusion, including congenital lung hypoplasia and bronchopulmonary dysplasia (BPD), and the remarkably limited lung capacity to recover from these defects or disorders. Therefore, understanding how to properly balance the self-renewal with differentiation of lung stem/

progenitor cells and its regulatory mechanisms could lead to designing innovative strategies for restoring normal morphogenesis of the lung and regenerating the *lung's surface for* gas diffusion.

The normal development of mammalian lungs requires a proper regulation of the cell division of epithelial stem/progenitor cells (Warburton et al. 2008, 2010). Different types of congenital lung defects and both developmental and birth disorders such as pulmonary hypoplasia and bronchopulmonary dysplasia may arise if the cell division of lung epithelial stem cells is not properly controlled. This will affect several biological processes in the lung that in turn affect the lung capacity for gas diffusion (Warburton et al. 2000, 2008, 2010; Shi et al. 2009). In humans, bronchopulmonary dysplasia (BPD) is an example of chronic pulmonary disease that severely affects both premature newborns and infants. Mechanical ventilation and the use of oxygen over *the long-term can cause* lung damages that will lead to bronchopulmonary dysplasia. Another well-known reparatory disease in humans is chronic obstructive pulmonary disease (COPD) that includes a group of respiratory tract diseases, which are characterized by airflow limitations or obstructions, and is mostly caused by smoking of tobacco or other well-knowing airway irritants. In some cases, chronic obstructive pulmonary disease may arise due to congenital defects. Furthermore, several lung studies have provided evidences that multiple cell fate decisions that are clearly linked to the cell division of undifferentiated epithelial stem cells can lead to a homogenous increase in the stem cell populations, through the asymmetric and/or symmetric cell divisions (Lu et al. 2008; Rawlins 2008; Berika et al., 2014; El-Hashash 2013, 2015; Elshahawy et al. 2016).

The balance between asymmetrical and symmetrical divisions and balancing self-renewal with differentiation of stem cells are required to preserve tissue homeostasis in different organs. Abnormal increases of self-renewal in tissue-specific stem cells can lead to tissue hyperplasia and/or tumorigenesis, while excessive stem cell differentiation may cause degeneration and/or aging of the tissue. In the lung, for example, identification of factors and molecular mechanisms that control the proper self-renewal-differentiation balance of lung-specific cell and progenitor cells can lead to new solutions to repair gas diffusion surfaces and restore lung morphogenesis. The asymmetric cell division is a required mechanism during development that enables the proper balance between stem cell self-renewal and differentiation and directs the temporal and spatial specifications of epithelial cell lineages in different body organs (Knoblich 2001; Yamashita et al. 2010; Vorhagen and Niessen 2014).

It is well established that stem and progenitor cells in different organs must strike a balance between self-renewal and differentiation to produce enough self-renewing daughters to progress through development and repair damage, without producing so many self-renewing daughter cells that can lead *to cancer development* (Vorhagen and Niessen 2014). Thus, identifying the molecular mechanisms that regulate the delicate balance between lung-specific stem cell self-renewal and differentiation is of fundamental importance not only to understanding lung development but also to exploiting the therapeutic potential of lung-specific stem cells.

The switch between the symmetric and asymmetric mode of cell division is a mechanism that maintains the self-renewing stem and progenitor cell compartments,

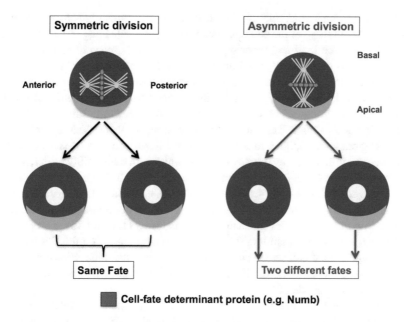

Fig. 5.1 Symmetric vs. asymmetric cell division in epithelial cells. Schematic depiction of a polarized dividing cell shows two modes of cell division. During symmetrical division, spindle orientation, and determinant protein (e.g., Numb) localization are not coordinated. Determinants segregate equally, giving rise to two equal (stem) cells. During asymmetric division, spindle orientation and determinant protein (e.g., Numb) localization are coordinated, giving rise to a differentiating cell and a stem cell. Thus, the difference in Numb (green) expression levels between two daughter cells mediates asymmetric cell division, whereas lack of Numb inheritance by both daughters will allow them to execute the stem cell self-renewal program by maintaining Notch1 activity and thus allowing symmetric cell division, as we reported in distal lung epithelial stem cells (El-Hashash and Warburton 2012; Berika et al. 2014)

which are also maintained by the balance between their capacities to self-renew and differentiate. The symmetric cell division generates two daughter cells that expand the stem cell populations if these cells have a stem cell fate. Conversely, the two daughter cells that gain a differentiation fate show less potential than the dividing mother cell (Fig. 5.1). This mode of cell division can rapidly produce tissue "effector" cells but also has the potential to deplete the stem cell pool (Molofsky et al. 2004; Yamashita et al. 2010). The asymmetric division is a mechanism of balancing self-renewal with differentiation of cells that results in a proper temporal and spatial specifications of various types of cell lineages throughout organ development. As opposed to the symmetrical mode of cell division, the asymmetric division of a mitotic stem or progenitor cell can produce two daughters with clearly distinct fates: one daughter cell that functions to maintain its stem cell identity and property and another daughter cell that can lose both stem cell property and function and acquire a differentiation fate (Fig. 5.1; Knoblich 2001). Details about the asymmetric divisions in lung stem and progenitor cells are described in Sect. 5.2.

Our careful understanding of the involved events and regulatory molecular mechanisms/pathways of the cell fate specification, and balancing self-renewal with different ion in the lung and other organs, is a key for humans in both health and diseases. This is particularly important because the far-reaching cell fate specification pathways and mechanisms in wide a variety of tissues make it particularly catastrophic when these pathways or molecular mechanisms fail to function normally, during tissue development, repair, and regeneration as well as during tissue homeostasis in the adult stages. This careful understanding is also critical for stem cells and their therapeutic applications because the current promise of stem cells and their effective utilization for *therapy* derive mainly from their ability to deftly navigate the multitude of multiple molecular mechanisms and signaling pathways that regulate the cell fate in a wide verity of cell types (Chen et al. 2013). Research in the last decades has uncovered many molecular components and signaling cascades that make up these molecular mechanisms and pathways function to specify the cell fate in diverse systems, including the atypical protein kinase Cζ (aPKCζ) that is an essential regulator of cell fate decisions in metazoans (Doe 2008; El-Hashash et al. 2011c; Homem and Knoblich 2012; Tepass 2012; El-Hashash and Warburton 2012; Drummond and Prehoda 2016).

In this chapter, we describe examples of the asymmetrical division in other organs, in which this mode of division is well studied (Sect. 5.1). We will also describe accumulated evidences and data on the asymmetrical cell division in the lung (Sect. 5.2) and discuss the similarity and difference of the asymmetrical stem cell division between the lung and other systems (Section 5.3).

5.1   Asymmetrical Divisions of Stem Cells in Different Systems

In different organ system, one of the most critical regulators of tissue architecture is *the oriented cell division* that is also important for tissue morphogenesis and homeostasis. The tightly regulated balance between the cell self-renewal/proliferation and differentiation is one of the important properties of different tissues to maintain and restore the homeostasis as well as to drive morphogenesis. Accumulated studies in different types of tissues show that the cell division orientation is closely related to the control of differentiation and leads to the production of daughters with a similar fate, which are produced by the symmetric cell division, or a differential fate that are produced by the asymmetric cell division (Fig. 5.2). These mechanisms enable the organism to generate remarkably diverse cell lineages from a small population of stem/progenitor cells. To be successful, a tight control of cell division orientation and/or the ratio of asymmetric and symmetric cell division is critical. Loss of the cell division orientation and/or an altered ratio between the asymmetric and symmetric cell division can result in tissue overgrowth, alteration of tissue architecture, and induction of aberrant differentiation. This loss and altered ratio have also been

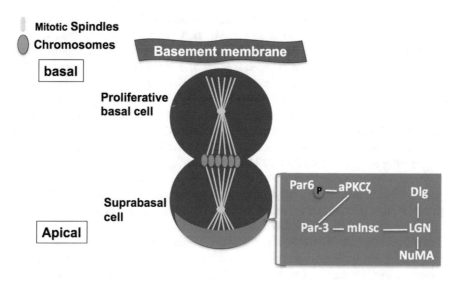

Fig. 5.2 Asymmetric cell division in mammalian epithelial cells. Schematic depiction of a polarized mammalian mother cell during mitosis (anaphase). Apical protein complexes are shown as a brown crescent. These apical protein complexes are important for both polarity establishment and spindle orientation in mammalian cells and are shown in a green box and described in the text. (Adapted from Berika et al. 2014)

linked to the development of cancer, multiple morphogenetic diseases, and aging. The establishment of a proper polarity axis is critically required for the oriented cell division. This can be achieved through several cell-specific intrinsic and/or extrinsic cues or signals. Major polarity-organization proteins, including the atypical PKC-Par-3-Par-6 polarity protein complex, are capable of translating such external and internal signals to drive proper cell polarity, as well as to regulate both the cell division orientations and cell fate in different types of epithelial cells in mammals (Fig. 5.2; Vorhagen and Niessen 2014).

The asymmetric mode of cell division represents a mechanism for multiple developmental and morphogenetic processes such as the cell fate decision, stem cell maintenance, and generation of differentiating daughter cells as well as the formation of the body axis. Proper cell polarity is essential for the asymmetrical cell division, and, consequently, a loss or defect of cell polarization may result in an abnormal increase of stem cell self-renewal that can lead to tumorigenesis.

A combination of specific intrinsic and extrinsic mechanisms and signals controls the asymmetric mode of cell division (Fig. 5.2). Several intrinsic mechanisms and signals are involved in this process such as the asymmetric cellular localization of the cell-specific intrinsic cell fate determinant molecules such as Numb, and/or cell-cell junctions, as well as the position of the cell within a specific environment or niche. These mechanisms may specify the cell polarity to direct the asymmetric mode of cell division (Fig. 5.2). The asymmetrical segregation and inheritance of the cell fate determinant molecules, including Numb, by one of the two daughters of a dividing cell requires a proper spindle orientation during mitosis, and other important coordination mechanisms and events. Conversely, the extrinsic mechanisms or factors are signals from neighboring cells that have a cell-cell communication with the dividing stem or progenitor cell. For example, the interactions between daughter cells themselves, or other surrounding cells, can regulate the specification of cell fate of these daughter cells in metazoans. In addition, direct genetic regulations of the asymmetrical cell division of dividing stem/progenitor cells and determination of their distinct fates are well reported in different systems (Betschinger and Knoblich 2004; Gómez-López et al. 2014; Berika et al. 2014; Dewey et al. 2015; Yang et al. 2015; Ibrahim and El-Hashash 2015).

In different organisms, the asymmetric segregation and inheritance of Numb, which inhibits Notch signaling activity, and/or activation of the Notch pathway are common mechanisms that direct the asymmetric cell division. The latter is the mode of division that is widely distributed in many different types of tissues throughout the animal kingdom. However, we still have *a rudimentary knowledge* of the regulatory molecular mechanisms and signal pathways of the asymmetric mode of cell division. Further studies are needed to identify and characterize the factors and molecular mechanisms that direct the asymmetric cell division. This includes the molecular mechanisms and signaling pathways that act to orient the mitotic spindles and to specify the cell fate. In this section, we describe examples of the asymmetrical mode of cell division that depends on the Notch signaling and are well studied in both mammalian and non-mammalian systems.

5.1.1   *Intestinal Stem and Progenitor Cells of* Drosophila

The intestinal stem and progenitor cells are well investigated compared to other types of stem cells. They, therefore, represent a model system for the asymmetric mode of cell division that is required for the maintenance of the intestinal epithelium. Moreover, the intestinal stem cells can generate several important cell types in the intestine such as polyploid enterocytes and entero-endocrine cells that produce hormones (Micchelli and Perrimon 2006; Ohlstein and Spradling 2006).

The Notch signaling mediates the asymmetrical division of intestinal stem and progenitor cells in the midgut of adult fruit flies (*Drosophila*; Ohlstein and Spradling 2007). This study has provided evidences that target genes for the Notch signaling in the daughter enteroblasts are activated by intestinal stem cell signals through the Notch ligand Delta. They have also found that the daughter enteroblasts exclusively activate the Notch signaling, and staining for the Notch ligand Delta is positive only in intestinal stem/progenitor cells, which directly contact the basement membrane (Ohlstein and Spradling 2007). In addition, the positive staining of the Notch ligand Delta is common in all clusters of stem cells (Ohlstein and Spradling 2007). Despite these remarkable findings by Ohlstein and Spradling (2007), the molecular mechanisms that block the Notch activity and thus facilitate the asymmetric division of intestinal stem cells still need further investigations. However, another study that involved the analysis of the mitotic spindle orientation of intestinal stem cells has shown that these cells do not divide randomly, because the daughter cell that retains its stem cell fate and identity was contacting the basement membrane, while the other daughter that gains a differentiation fate is displaced to eventually form an enteroblast (Toledano and Jones 2009). Further investigations are still required to determine how the orientation of mitotic spindle fibers is controlled in the intestinal stem cells.

5.1.2   Hematopoietic Stem and Progenitor Cells

Like the studies of intestinal stem and progenitor cells in *Drosophila,* the importance of the Notch signaling is also implied by several investigations that have used the methods of blocking differentiation in *Drosophila* intestinal stem and progenitor cells to characterize the hematopoietic stem cell fate (Duncan et al. 2005; Wu et al. 2007). In these studies, a Notch GFP reporter was used to enrich for murine hematopoietic stem and progenitor cells. Interestingly, the Notch-specific reporter strain that was used by Duncan and Wu and their co-workers showed that GFP+ cells had around 40–60% hematopoietic stem cells, whereas GFP expression was significantly reduced in differentiating precursor cells (Duncan et al. 2005; Wu et al. 2007). This strongly supports the significance of both intracellular cues and extrinsic factors in the regulation of cell division of hematopoietic stem and progenitor cells. In addition, various types of oncogenic translocations on chromosomes, including BCR-ABL, influence the hematopoietic stem cell survival and the either pattern of cell division or proliferation (Duncan et al. 2005; Wu et al. 2007). More investigations are still required to understand the signaling pathways and both molecular and functional mechanisms that regulate the asymmetric cell division by these factors.

5.1.3   Muscle Stem and Progenitor Cells

In various organisms, satellite cells that are located close to myofibers localize underneath the basement membrane and functionally act as stem/progenitor cells of muscle tissue. Satellite cells are quiescent but are capable of entering the cell cycle if the tissue is injured. These cells are, therefore, important in the maintenance of myoblast cell populations during muscle development, regeneration, and repair after injury. During the processes of muscle growth, repair, and regeneration, daughter cells of the dividing muscle lineage are characterized by an asymmetrical segregation of the younger and older (immortal) strands of DNA (Cairns 1975; Shinin et al. 2006; Conboy et al. 2007).

Furthermore, many in vitro and in vivo studies on muscle stem cells show evidences of the upregulation of several differentiation genes, and the activity of major cell fate determination proteins, including Numb, within the progeny of dividing cells, which support the existence of the asymmetric mode of cell division among muscle stem cells (Conboy and Rando 2002; Shinin et al. 2006; Kuang et al. 2007). However, further investigations are still needed to support the importance of the asymmetric mode of cell division in muscle cell repair and regeneration. Furthermore, more studies are still needed to both identify and characterize the molecular and functional mechanisms as well as signaling pathways that govern muscle satellite cell divisions in response to many factors in their environment, which can determine the daughter cell fates.

5.1.4   Epidermal Stem and Progenitor Cells

The skin provides a protective barrier of the body against its environment and represents the largest organ of the body. Weathering the effects of the outside world requires this organ to have a rapid and efficient repair mechanism. Therefore, the asymmetric division in the skin is critical for the maintenance of the skin barrier. The importance of the asymmetric cell division for the skin is evidenced by the conservation of the molecular mechanisms, factors, and signaling pathways that are responsible for the asymmetric cell division throughout evolution.

Several in vivo studies on murine embryonic skin cells and other studies using skin cells that were grown in vitro have provided evidences of both types of cell divisions, symmetric and asymmetric, in the mammalian epidermal stem/progenitor cells. Indeed, the asymmetric mode of cell division within the basal layer of the esophageal epithelium was reported *almost two decades ago* (Seery and Watt, 2000). In addition, it is well established that the stem cells of the epidermis' basal layer can divide asymmetrically to form the stratified epithelium or symmetrically to generate more stem cells (Smart 1970; Lechler and Fuchs 2005). The dividing epidermal stem or progenitor cell gives rise into two types of daughter cells by the asymmetric mode of cell division. One daughter can proliferate and localize in the basal layer and is contacted to the basolateral membrane. The other daughter cell is detached and becomes part of the extended supra-basal layer of the skin (Lechler and Fuchs 2005). At later developmental stages, the supra-basal cells halt division and start to differentiate to produce the characteristic skin barrier layer (Fuchs and Raghavan 2002).

In their remarkable study, Lechler and Fuchs (2005) have reported that the mitotic spindle fibers of asymmetrically dividing cells are oriented perpendicularly to the basement membrane during the process of skin stratification. Both integrins and cadherins are required for multiple biological and cellular processes in epidermal stem cells and were found as critical adhesion molecules for the physical attachment of basal stem and progenitor cells to the basement membrane (Lechler and Fuchs 2005). Integrins and cadherins are, therefore, essential for the alignment of the mitotic spindles (Lechler and Fuchs 2005). Moreover, both integrins and growth factor receptors can affect the behavior of stem cells (Lechler and Fuchs 2005). Furthermore, the analysis of the perpendicular cell division shows that this type of cell division acts as a natural mechanism that unequally partitions signaling molecules of the basement membrane into the issuing daughter cells (Lechler and Fuchs 2005).

In the asymmetric mode of cell division, mitotically dividing cells have characteristic perpendicular to spindle fibers and an apically localized LGN polarity/spindle orientation regulatory protein, which is the Pins ortholog in mammals. The LGN protein can bind to other polarity and/or spindle orientation regulatory proteins such as mInsc and Par3 in basal cells at their apical cortex. The atypical PKC (aPKCζ) is another polarity protein that is apically localized in the basal cells (Lechler and Fuchs 2005). In addition, LGN can bind to the spindle orientation regulatory protein NuMA, which is the Mud ortholog and tethers the mitotic spindles to the cell poles (Du et al. 2001). Furthermore, both integrins and cadherins cell adhesion molecules play important roles in the polarized localization of polarity proteins such as NuMA-dynactin, aPKCζ, Par3-LGN-Inscuteable polarity complex at the apical cell side, as well as in the alignment of mitotic spindles to basal stem/progenitor cells (Lechler and Fuchs 2005). Notably, the major functional roles of apical polarity proteins in epithelial stem cells of mammals are to determine the position of the mitotic spindles and establish the cell polarization, rather than cell fate determination (reviewed in Macara 2004a, b; Lechler and Fuchs 2005; Suzuki and Ohno 2006; Shin et al. 2007; Knoblich 2010; Yamashita et al. 2010).

Transcription factors are also important for the cell behavior and the asymmetric division of stem cells of the epidermis. For example, p63 promotes the proliferation of epidermal cells and plays an important role in the asymmetrical division of mammalian epidermal stem/progenitor cells (Mills et al. 1999; Yang et al. 1999). Deletion of *p63* gene promotes symmetrical divisions in basal cells, suggesting the requirement of *p63* for both cell stratification and the choice of cell division type in these cells (Lechler and Fuchs 2005; Senoo et al. 2007).

The Notch signaling pathway is a critical regulator of the asymmetrical cell division in mammalian epithelial stem cells, likes *Drosophila* neuroblasts and other stem cell types. Blanpain and co-workers have shown that the Notch intracellular domain (NICD) is utilized by the supra-basal cells and promotes the cell differentiation (Blanpain et al. 2006). Other studies show that the cellular localization of Numb to the basolateral cell side is dependent on its phosphorylation and activation status that is controlled by aPKCζ (Smith et al. 2007). Numb functions to inhibit the activity of Notch signaling to regulate the cell fate (Smith et al. 2007).

Despite the progress in our understanding of the cell behavior and mechanisms of cell division in the skin epithelial cells, further investigations are still required to determine the regulatory molecular mechanisms and signaling pathways of the asymmetrical cell division in stem/progenitor cells of mammalian epidermis during both developmental and adult stages, as well as during tissue repair and regeneration.

5.1.5   Neural Stem and Progenitor Cells

The cell division mode is well studied in neural stem cells of mammals and *Drosophila*. Both symmetric and asymmetric modes of the cell division occur throughout different developmental stages of mammalian neural stem cells. During development, the symmetric cell divisions occur to possibly increase the neural stem cell pools, after which the asymmetric mode of cell division takes place to generate more differentiating neuronal populations. The asymmetrical cell division has been detected in the ventricular zone that is localized in mammalian cerebral cortex, as well as in the neuroepithelium of the vertebrate retina (Gonczy 2008; Neumüller and Knoblich 2009; Yamashita et al. 2010).

In the nervous system of vertebrates, the orientation of spindle fibers is controlled by LGN, mInsc, NuMA, and other molecules or factors, which have a conserved role and, therefore, exist in other types of cells and systems. In addition, the Notch signaling pathway can regulate the cell fate in the nervous system of vertebrates (Chenn and McConnell 1995; Zhong et al. 1996, 1997; Petersen et al. 2004). Like many tissue types, the Notch signaling activity is suppressed by Numb in the nervous system during the determination of cell fate. In addition, Numb may be also involved in other molecular mechanisms that determine the fate of nerve cells (Rasin et al. 2007; Zhou et al. 2007).

Many research studies on other systems show that the orientation of the mitotic spindles can be used for the prediction of the mode of cell division (Sanada and Tsai 2005; Zigman et al. 2005; Morin et al. 2007; Konno et al. 2008). However, it is not yet clear whether there is a correlation between the specific mitotic spindles fibers' orientation and the determination of cell fate of the developing nervous system in vertebrates. For example, a recent study has shown that altering the activity of LGN (a Gα-binding protein) that controls the mitotic spindle orientation can lead to random spindle orientations in the spinal cord neuroepithelium. However, this disruption did not apparently disrupt the daughter cell fate (Morin et al. 2007). In addition, a reduction in the activity level of mInsc, which is another major regulator of the orientation of mitotic spindle fibers, results in increased of stem cell numbers (Zigman et al. 2005). However, this increase was accompanied with some neuronal defects probably because of the disrupted orientation of the mitotic spindle fibers (Zigman et al. 2005).

Calcium ions (Ca2+) are another important factor in the neuronal cell fate. Several studies have described the important role of Ca2+ in tissue homeostasis and both the signaling potential of Ca2+ and its functional role in human embryonic stem/progenitor cell differentiation into a neuronal phenotype (reviewed in Forostyak et al. 2013). This stem cell differentiation into the neuronal phenotype is correlated with Ca2+ signaling activity (Forostyak et al. 2013). Notably, human embryonic stem/progenitor cells may transiently attain an actively operating period of Ca2+ signaling with time in culture, which is apparently important for their differentiation into the neuronal phenotype (Forostyak et al. 2013).

5.2   Asymmetrical Divisions of the Lung Epithelial Stem and Progenitor Cells

The tight regulation of stem and progenitor cells is important for proper lung development and function in mammals (Warburton 2008; Warburton et al. 2010; El-Hashash 2013, 2015; Berika et al. 2014, 2016; Elshahawy et al. 2016; Ku and El-Hashash 2016). Several congenital and developmental lung defects or disorders, including severe lung hypoplasia and bronchopulmonary dysplasia, are caused by an improper stem cell regulation or behavior. Some of these lung defects or disorders can significantly alter vital life and biological processes that influence the capacity of gas diffusion in the lung and, therefore, they are lethal (Warburton et al. 2000, 2008, 2010; Shi et al. 2009). Determination of signals and molecular mechanisms that are required to properly establish the balance of cell proliferation (or self-renewal) with differentiation of lung stem/progenitor cells is a key because it may help identifying novel solutions or therapies for restoring lung morphogenesis and new methods for repair and regenerate of the lung after injury.

The homogeneity of stem cells in the lung is a result of controlled symmetric and asymmetric cell divisions (Lu et al. 2008; Rawlins 2008). Careful observations of the orientation of spindle fibers or segregation and inheritance of key determinant molecules of the cell fate such as Numb by the two daughter cells of a dividing stem/progenitor cell can help in the differentiation between symmetrical and asymmetrical cell divisions (Huttner and Kosodo 2005; Morrison and Kimble, 2006; Wang et al. 2009; El-Hashash and Warburton 2011, 2012; Berika et al. 2014; Ibrahim and El-Hashash 2015).

The asymmetric mode of cell division is an essential mechanism during both development and morphogenesis because it leads to the balance between proliferative and differentiated cell populations. In addition, the asymmetric mode of cell division corrects the spatial-temporal specifications of the cell lineages in epithelial cells (reviewed in detail in Yamashita et al. 2010; Knoblich 2001; Gómez-López et al. 2014; El-Hashash 2015; Elshahawy et al. 2016). Despite the importance of the asymmetric mode of cell division, and its closely related processes such as the cell polarization in lung-specific epithelial stem and progenitor cells, we still have a rudimentary understanding of the cellular mechanisms and regulatory signaling pathways.

The asymmetric mode of cell division is an essential mechanism during both development and morphogenesis because it leads to the balance between proliferative and differentiated cell populations. Moreover, the asymmetric mode of cell division corrects both the spatial and temporal specifications of cell lineages in epithelial cells (see Knoblich 2001; Yamashita et al. 2010, Gómez-López et al. 2014; El-Hashash 2015; Elshahawy et al. 2016 for detailed reviews). Despite the importance of the asymmetric mode of cell division, and its closely related processes such as the cell polarization in lung epithelial stem and progenitor cells, little is known about their cellular mechanisms and regulatory signaling pathways.

Both extrinsic and intrinsic cell fate determinants are important in the asymmetrical stem cell division. The cell microenvironment is an example of the extrinsic fate determinants. If daughter cells that were produced from one dividing progenitor cell are placed in two different microenvironments, they will undergo different cell fates. The cytoplasmic fate determinants such as Numb protein are good examples of the intrinsic determinants of cell fate. The asymmetrical cell division in different systems depends largely on the preferential distribution, segregation, and inheritance of the intrinsic determinants of cell fate, as reported in mammalian and *Drosophila* epithelial cells (Cayouette and Raff 2002; Betschinger and Knoblich 2004). In the lung, the asymmetrical divisions were reported in stem and progenitor cells of the distal lung epithelium and were evident by the correlation between perpendicular asymmetrical divisions and Numb asymmetrical inheritance in these dividing cells (Fig. 5.1; El-Hashash et al. 2011c; El-Hashash and Warburton 2012). In addition, the knockdown of Numb in murine lung epithelial cells 15 (MLE15) leads to more cells that express the stem/progenitor cell markers Id2 and Sox9 in culture, supporting the function of Numb in lung stem/progenitor cell fate determination (El-Hashash et al. 2011c; El-Hashash and Warburton 2012). The role of the cell fate determinant molecules such as Numb in the asymmetrical division and behavior of stem and progenitor cells in the lung is described in Sect. 5.2.1.

The asymmetrical cell division in the distal lung epithelium plays important functional roles in the maintenance of epithelial stem/progenitor cells and formation of the lung by balancing proliferation/self-renewal with differentiation of the lung-specific stem cells. This can also lead to the maintenance of both differentiated and stem cell populations (El-Hashash et al. 2011c; El-Hashash and Warburton 2011, 2012) that is supported by several experimental evidences. For example, epithelial stem and progenitor cells in the distal embryonic lung are polarized and highly mitotic with characteristic perpendicular to cell divisions (Figs. 5.3 and 5.4; El-Hashash et al. 2011c; El-Hashash and Warburton 2011, 2012). The perpendicular division is highly correlated with the asymmetrical division in the epithelium, as is the case in different types of mammalian epithelial cells that have been reported to undergo the asymmetric divisions by changing the orientation of their mitotic spindle fibers from a parallel position into a perpendicular position (Lechler and Fuchs 2005). These are consistent with the asymmetrical, apical, and polarized localizations of several polarity proteins such as NuMA, mouse Inscuteable (mInsc), and LGN (Gpsm2) that control the orientation of mitotic spindles in mitotically dividing distal stem and progenitor cells of the lung epithelium during embryogenesis (Figs. 5.4 and 5.5; El-Hashash et al. 2011c; El-Hashash and Warburton 2011), as well as in other types of epithelial cells (Lechler and Fuchs 2005). In addition, the interference with these polarity and mitotic spindle regulatory proteins in vitro can lead to random distributions of spindle fibers during mitosis, loss of the balance between self-renewal and differentiation, and changes in the cell fate of lung epithelial cells grown in culture (El-Hashash et al. 2011c; El-Hashash and Warburton 2011, 2012). Furthermore, intensive analyses of the cadherin hole in the lung

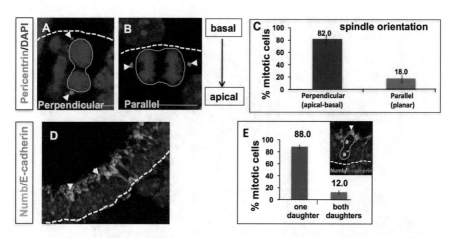

Fig. 5.3 Perpendicular as well as asymmetric cell division in embryonic lung distal epithelial stem and progenitor cells. Panels (**A–E**) which are adapted from El-Hashash and Warburton (2011) show evidences of perpendicular and asymmetric cell division in distal epithelial stem/progenitor cells of embryonic lungs (E14). (**A, B**) Most mitotic cells in the distal epithelium divide perpendicularly, as represented by the perpendicular orientation of pericentrin-stained centrosomes (**A**, arrowheads, arrows) relative to the basement membrane (dashed line; **A–B**), while only a few mitotic cells have their centrosomes aligned parallel to the basement membrane (**B**, arrowheads). (**C**) Quantitation of spindle orientations, expressed as a percentage of all divisions in the distal epithelium from the experiments shown in **A, B** ($n = 48$). (**D**) Expression of the cell fate determinant Numb in lung distal epithelial stem/progenitor cells, as shown by immunofluorescence. Note polarized apical localization of Numb (arrowheads), which is a key factor in asymmetric cell division, relative to collagen IV-stained basement membrane (dashed line). (**E**) Quantification of late mitotic distal epithelial cells, with Numb inherited by one (inset in **E**) or both daughter cells at E14. This is expressed as a percentage of all divisions in the distal epithelium ($n = 42$)

epithelium provide further evidences that the asymmetric division is common in distal epithelial stem and progenitor cells (El-Hashash and Warburton 2012).

In different organ systems, the epithelium has a characteristic apical-basal cell polarity. In addition, the epithelial cell has a distinct morphology and shape of which a slight deviation from normalcy can result in an asymmetric distribution and localization of the apical plasma membrane along with the adherent junctions to the daughters of a mitotically dividing cell (Nelson 2003a, b; Kosodo et al. 2004). The role of cell polarity in the determination of the mode and outcome (asymmetric vs. symmetric) of stem/progenitor cell division is described in Sect. 5.2.2.

In summary, the characterization of the asymmetric mode of cell division and its regulatory mechanisms and the investigation of lung epithelial stem cell behavior and how these stem cells control and balance their different fates have been partially uncovered in the lung. However, further research is still needed in these research areas since it will help identify novel targets to prevent and/or rescue various types of fatal or serious lung diseases. It can also help in devising novel methods for lung repair and regeneration. In addition, understanding the mechanisms of balancing proliferation/self-renewal with differentiation of stem/progenitor cells can help to

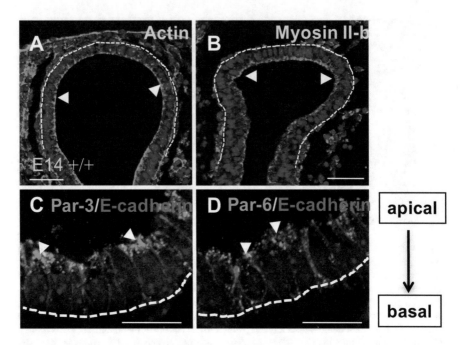

Fig. 5.4 Cell polarity in embryonic lung distal epithelial stem and progenitor cells. Panels (**A–D**), which are adapted from El-Hashash and Warburton (2011), show evidences of cell polarity in distal epithelial stem and stem and progenitor cells of embryonic lungs (E14). (**A–D**) Immunofluorescence at E14 shows strong signals for actin (**A**), myosin II-b (**B**), as well as the polarity proteins: Par-3 (**C**) and Par-6 (**D**) proteins at the apical side of distal epithelial progenitor cells (arrowheads). Dashed line represents the collagen IV-stained basement membrane. Scale bars. 50 mm

develop research methods that utilize the remarkable ability of these stem cells to repair and regenerate the damaged and/or diseased lung.

5.2.1   Numb as a Key Regulator of Asymmetric Cell Division and Cell Fate in the Lung

Numb is a cell fate determinant that play a major role in both asymmetric and symmetric cell division (reviewed in Morrison and Kimble 2006; Knoblich 2001, 2010; Zimdahl et al. 2014 Liu et al. 2015). In mammals, four isoforms of Numb protein exist, whereas only one form of Numb is expressed in *Drosophila*. Numb protein is generally well studied in *Drosophila* than in mammals (reviewed in Gulino et al. 2010). The gene, *NUMB*, encodes Numb protein that plays an important function in the regulation of the cell fate decisions in different tissues and systems during organogenesis such as the central and peripheral nervous systems in vertebrates and invertebrates animals (Pece et al. 2011; Gulino et al. 2010). This function and

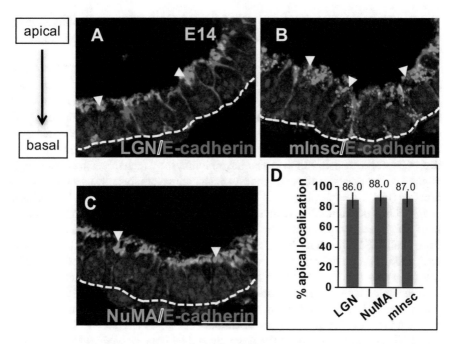

Fig. 5.5 Polarized localization of spindle orientation-regulatory proteins in the lung distal epithelial stem and progenitor cells. Panels (**A–D**), which are adapted from El-Hashash and Warburton (2011), show evidences of cell polarity in distal epithelial stem and progenitor cells of embryonic lungs (E14) based on the expression pattern of spindle orientation-regulatory proteins that are also polarity proteins: LGN, mInsc, and NuMA. The expression of LGN, mInsc, and NuMA proteins in lung distal epithelial stem and progenitor cells is shown by immunofluorescence (**A–C**). Note polarized apical localization of these proteins (**A–C**, arrowheads) relative to collagen IV-stained basement membrane (dashed line). (**D**) Quantitation of the apical localization of proteins shown in A–C, which is expressed as a percentage of all cells in the distal epithelium ($n = 87$). Scale bars. 50 mm

others, including the control of the asymmetric mode of cell division by enabling differential cell fate specification of dividing cells, are intensively studied during the development of nervous system in *Drosophila* and mammals (Betschinger and Knoblich 2004).

The asymmetric mode of cell division is dependent on the preferential segregation of Numb protein or other intrinsic cell fate determinants into one of the two sibling daughters during the cell division of *Drosophila* and mammalian epithelial cells (Betschinger and Knoblich 2004). Numb has a characteristic pattern of distribution in the cell, in which it localized asymmetrically during mitosis. Numb protein, therefore, has a polarized localization at one side of the dividing stem cell. Then, Numb segregates to only one daughter cell asymmetrically, in which it acts to determine the fate of this cell intrinsically (Gulino et al. 2010). This pattern of Numb cellular localization and segregation in dividing cells can enable Numb to define the polarity axis, which determines the orientation of the apical-basal cell division plane in these cells. It also allows a rapid switch from cell proliferation to cell diversification in different tissues and systems (Betschinger and Knoblich 2004).

The cell fate determinant Numb has a *phosphotyrosine*-binding (PTB) domain and a C-terminal region, which has an Eps15 homology (EH) domain containing proteins and a conserved binding motif for α-Adaptin. Numb protein is uniformly expressed in the cell cytoplasm during the interphase but is localized to one cell side during cell division. In dividing cells, Numb is asymmetrically segregated and inherited by one daughter cell only, enabling this cell to adopt a different fate from that of its sibling. Since Numb is a Notch inhibitor, the daughter cell receiving high Numb levels suppresses the extrinsic Notch signaling activity and has a differentiation fate (Fig. 5.1; Betschinger and Knoblich 2004; Alexson et al. 2006, El-Hashash et al. 2011c). In contrast, the daughter cell with low levels of Numb normally maintains a high activity of Notch signaling and, thus, acquires a stem cell fate (Fig. 5.1; Frise et al. 1996; Guo et al. 1996; Frise et al. 1996; Juven-Gershon et al. 1998; Yan et al. 2008). During neurogenesis, Numb protein normally has a polarized cell localization at one side of the mitotic mother cell, such that it is segregated and then inherited selectively by one daughter cell. This mode of asymmetric segregation and inheritance of Numb protein allows the daughter cell that contains Numb to have a different fate than the other daughter cell that does not inherit Numb (Gulino et al. 2010). This asymmetric mode of segregation and inheritance enables Numb to play a major functional role in many biological processes, including the binary cell fate decisions, tumorigenesis, and the migration of neural progenitor cells (Gulino et al. 2010).

Proper control of epithelial stem and progenitor cells is critical for normal formation, development, and function of the mammalian lung (Warburton 2008; Warburton et al. 2000, 2010; Berika et al. 2014; Ibrahim and El-Hashash 2015). A significant deficiency of lung- specific stem or progenitor cells may lead to many lethal defects of gas diffusion capacity in the lung, including the common congenital forms of both pulmonary hypoplasia and bronchopulmonary dysplasia (BPD) as well as the limited capacity of the lung to recover from these defects (Shi et al. 2009; Warburton et al. 2008, 2010). Therefore, studies that aim to understand how the proper balance between different cell fates (cell proliferation, differentiation, and apoptosis) of lung-specific stem cells is achieved could lead to the discovery of many innovative solutions to restoring normal morphogenesis of the lung and possibly regeneration of the gas diffusion surface. Many recent studies have identified the asymmetric mode of cell division as an important mechanism of balancing cell self-renewal and differentiation and maintaining the correct spatiotemporal specification of cell lineages in epithelial cell types of different organs (Knoblich 2001, 2010; Berika et al. 2014; Chen et al. 2016; Fu et al. 2017; Daynac and Petritsch 2017).

The asymmetric vs. symmetric cell division and Numb activities and functions are well-characterized in *Drosophila* and mammalian nervous systems during organogenesis (Knoblich 2001, 2010; Betschinger and Knoblich 2004). In the lung, our studies have demonstrated the expression pattern and function of the cell fate determinant Numb in dividing distal epithelial stem and progenitor cells. Both apical- basal polarity and perpendicular cell division are common in distal lung epithelial stem and progenitor cells (Fig. 5.3; El-Hashash and Warburton 2011, 2012; El-Hashash et al. 2011c). The cell fate determinant Numb has a characteristic polarized apical cell localization in distal lung epithelial stem and progenitor cells (Fig. 5.3; El-Hashash and Warburton 2011, 2012). During early lung development, the expression levels of Numb protein are weak in distal murine lung epithelial stem and progenitor cells (at E11.5–12.5 in mouse embryos). The expression level of Numb protein is enhanced in the distal rather than proximal epithelial cells during later developmental stages of murine lung (El-Hashash and Warburton 2012). Another Numb-related protein, α-Adaptin, is an endocytic protein that is important for Numb functions in mediating the asymmetric cell division (Berdnik et al. 2002). Both Numb and Numb-associated signal α-Adaptin proteins have a characteristic polarized and asymmetric distribution at the apical side of distal epithelial stem/ progenitor cells of the developing murine lung (El-Hashash and Warburton 2012). During mitotic cell division, Numb protein is asymmetrically segregated to and inherited by one daughter in most mitotically dividing distal epithelial stem/ progenitor cells in the lung (Fig. 5.3; El-Hashash and Warburton 2011, 2012 El-Hashash et al. 2011c). The more perpendicular/asymmetric the cell division is, the more likely it is to segregate Numb preferentially to one daughter cell in mitotic distal lung epithelial stem cells; it was suggested that asymmetric division is the mode of cell division in these cell types during lung embryogenesis (El-Hashash and Warburton 2011, 2012).

Studies on the mechanisms of Numb functions have indicated that it can control the cell fate by inhibiting the activity of Notch signaling through a characteristic polarized receptor-mediated endocytosis. It is well established in different systems that the Notch signaling can promote the identity of stem cells at the expense of differentiated cell phenotypes. Numb protein can act as a linker between both Notch and α-Adaptin that is a facilitator of Notch endocytosis (Giebel and Wodarz 2012). In addition, Numb has a characteristic pattern of localization and segregation before and during mitotic cell division (Frise et al. 1996; Guo et al. 1996; Juven-Gershon et al. 1998; Yan et al. 2008; Berika et al. 2014). Numb is uniformly expressed in the cytoplasm during interphase, and then, it has an asymmetric localization pattern in the mitotically dividing cell. Hence, Numb is inherited by only one daughter cell, enabling this cell to adopt a different fate from that of its sibling. The cell with low Numb levels can, therefore, maintain a high level of Notch activity and thus acquire a stem cell fate, whereas the cell receiving high Numb activity levels suppresses the activities of extrinsic Notch signaling and differentiates (Frise et al. 1996; Guo et al. 1996; Juven-Gershon et al. 1998; Yan et al. 2008; Berika et al. 2014). Predictably, deletion of *Numb* gene in culture can lead to an increase of the Notch signaling activity, in combination with an enhancement of the number of lung epithelial stem cells growing in vitro (El-Hashash and Warburton 2012). Numb protein may, therefore, have a conserved function in the control of Notch signaling activity and in cell fate determinations in lung epithelial stem cells (El-Hashash and Warburton 2012). These studies are to be commended for showing, for the first time, that Numb is a crucial regulator of the lung stem cell fate that is an important step forward in our understanding of lung development, repair, and regeneration (El-Hashash et al. 2011c; El-Hashash and Warburton 2011, 2012). However, more investigations are still needed to identify the mechanisms underlying Numb functional activities in lung stem cells.

5.2.2   Cell Polarity as a Major Regulator of Asymmetric Cell Division and Cell Fate

Accumulated experimental data strongly support the hypothesis that the establishment of polarity during mitosis is another major regulator of both the mode and outcome (asymmetric vs. symmetric) of stem/progenitor cell division. The epithelial cells are characterized by their apical-basal polarity. The cell polarity is recognized by an asymmetrical distribution of specific cellular constituents within the cell. The proper establishment of the cell polarization is crucial for the asymmetrical cell division as well as for multiple cellular processes such as the cell specification and migration. In addition, the proper cell polarization plays an important role in several cellular processes and mechanisms that organize and integrate the multitudes of molecular signals, which affect various cell decisions related to the cell fate, differentiation, orientation, and proliferation. They also affect the interactions with other cell types and the surrounding extracellular matrix. Furthermore, the cell polarity is essential for the architecture and function of epithelial cells in different tissue types (Wodarz 2002; Nelson 2003a, b; Florian and Geiger 2010; Roignot et al. 2013; Vorhagen and Niessen 2014; Dewey et al. 2015; Ajduk and Zernicka-Goetz 2016).

Establishing a proper epithelial cell polarization involves the intervention of several fundamental biological cell processes and molecular mechanisms. Therefore, to understand the complex process of building various types of epithelial tissues during development, repair, and regeneration, it is necessary to identify and characterize the underlying events and molecular mechanisms that govern the epithelial cell polarization. In recent decades, several research methods and tools have been developed to identify and characterize the regulatory molecular mechanisms of the establishment and maintenance of the architecture of both simple and higher-order epithelial tissues. These developed research methods and tools have also helped in both understanding the dynamic remodelling of the epithelial cell polarization, which usually takes place during the development and morphogenesis of different types of epithelial organs (Roignot et al. 2013). Recent research progresses have been achieved on how cell polarity cues, cell-to-cell communication, and cell cytoskeleton interact to control the division plan orientations during the embryogenesis of *Drosophila*, *Caenorhabditis elegans*, and mice (Ajduk and Zernicka-Goetz 2016). In addition, recent research has improved our understanding of the functions of Par polarity protein complex in the cell division orientation during the cleavage of mammalian embryos (Ajduk and Zernicka-Goetz 2016).

The asymmetrical distribution of some components within the cell plays a major role in the establishment of cell polarization. In addition, proper cell polarization and orientation of the mitotic spindle are critical for balancing self-renewal with differentiation in various stem and progenitor cell types and can affect multiple physiological processes, such as the epithelial differentiation, repair, and regeneration as well as tissue branching morphogenesis. In addition, a proper cell polarity helps to organize and integrate multiple and complex molecular signals within the cell (Vorhagen and Niessen 2014; Dewey et al. 2015; Yang et al. 2015; Overeem et al. 2015; Tuncay and Ebnet 2016).

One of the major regulators of epithelial cell polarization is Par polarity complex proteins. The *Par* genes were discovered by several genetic screens' studies on the regulators of cytoplasmic partitioning during the early embryogenesis of *C. elegans*. The *Par* genes encode six different proteins that are critical regulators of both cell polarization and asymmetrical division in various types of animal epithelial tissues. Most Par proteins are polarized with an asymmetric localization at the apical cell side. They form a physical complex, Par polarity complex, with one another inside the epithelial cell (Goldstein and Macara 2007).

Proper mitotic spindle orientation is another factor affecting both the cell polarity and mode of cell division. Even a slight deviation of the spindle orientation in a dividing epithelial cell may lead to a loss of symmetrical division and a switch to asymmetrical cell division (Nelson 2003a, b; Kosodo et al. 2004; Lechler and Fuchs 2005; Peng and Axelrod 2012). It is well established that in mitotically dividing epithelial cells of different tissues, the "cadherin hole" can be seen as unstained segments in the cell surface by immunostaining of E-cadherin that exists in the lateral epithelial cell plasma membrane and the apico-lateral junction complex (Woods et al. 1997; Kosodo et al. 2004; El-Hashash and Warburton 2012). Remarkably, the cleavage plane orientation relative to the cell surface' cadherin hole can provide a prediction for distribution of mitotic cell's plasma membrane to daughter cells, and whether it is asymmetrical or symmetrical distribution (Kosodo et al. 2004).

In the lung, the proper establishment of cell polarity and, thus, cell fate is critical for the alveolar development and function, but it is partially characterized in the lung. The formation of a selective and permeable monolayer, in which the contact between cells can provide several spatial cues that promote the generation of a proper cell polarity, is a prerequisite for the effective respiratory functions of the alveolar surface (Nelson 2003a, b; Boitano et al. 2004; (Vorhagen and Niessen 2014). The cell polarity is, therefore, crucial for both proper function and specification of different epithelial cell lineages in the lung.

Stem and progenitor cells of the distal epithelium of embryonic lung are polarized and highly mitotic (Figs. 5.4 and 5.5; El-Hashash and Warburton 2011, 2012; El-Hashash et al. 2011c; Ibrahim and El-Hashash 2015). In addition, the establishment of a proper cell polarization is important for both perpendicular divisions and balancing self-renewal with differentiation in lung stem and progenitor cells (Figs. 5.3 and 5.4; El-Hashash and Warburton 2011, 2012; El-Hashash et al. 2011c; Ibrahim and El-Hashash 2015). A disruption in the cell polarization leads to loss of the balance between lung epithelial cell self-renewal and differentiation in culture (El-Hashash and Warburton 2011; El-Hashash et al. 2011c).

The correlation between the asymmetrical divisions and perpendicular divisions was reported in the epithelial cells of other mammalian tissues (Lechler and Fuchs 2005; Peng and Axelrod 2012). This correlation was also reported in lung stem and progenitor cells by the detection of the asymmetric localization of proteins that control the perpendicular orientation of spindle fibers, including G-protein signaling modulator 2 (GPSM2), mouse Inscuteable (mInsc), and nuclear mitotic apparatus (NuMA) polarity proteins, in the mitotic distal lung epithelial stem cells that show asymmetric divisions (Figs. 5.4 and 5.5; El-Hashash and Warburton 2011, 2012; El-Hashash et al. 2011c). Notably, the disruption of asymmetric divisions was closely associated with the loss of both polarized and apical cell side localization of the proteins controlling the perpendicular spindle orientation in lung epithelial cells in vivo and in culture (El-Hashash and Warburton 2011; El-Hashash et al. 2011c).

5.2.3   Protein Phosphatases Functions in the Asymmetric Cell Division and Cell Fate in the Lung

Eya1 protein phosphatase is intensively expressed in lung distal epithelial stem/progenitor cells and plays a crucial role in the establishment of proper cell polarity and orientation of the mitotic spindle fibers in vitro and in vivo (El-Hashash et al. 2011c). Eya1 phosphatase controls the cell polarity, and thus cell fate by targeting and regulating the activity of aPKCζ, which is a well-known regulator of both the perpendicular divisions and the asymmetrical Numb inheritance in dividing cells (El-Hashash et al. 2011c; Tepass 2012; Drummond and Prehoda 2016), in lung distal epithelial stem and progenitor cells. Active aPKCζ, in turn, controls the cell polarity by forming a complex with Par-3 and Par-6 polarity proteins (El-Hashash et al. 2011c).

Eya1 also controls the fate of stem and progenitor cells in distal lung epithelium by regulating the phosphorylation and activity of Numb, which is a Notch inhibitor, and consequently Numb asymmetric inherence in dividing cells (El-Hashash et al. 2011c). During mitosis, Eya1 phosphatase functions as an inducer of both the perpendicular cell division and Numb asymmetric segregation and inheritance by one daughter cell of the dividing distal lung epithelial stem cells, probably by controlling the activity and phosphorylation levels of protein kinase C (aPKCζ; El-Hashash et al. 2011c). Eya1 deletion in vivo or in culture results in an uncontrolled increase of aPKCζ phosphorylation and activity that inhibits the polarized apical localization of Par polarity proteins, which perturbs the cell polarity (El-Hashash et al. 2011c). If *Eya1* is deleted in vivo or in culture, aPKCζ-controlled Numb segregates to and is inherited by both daughters, leading to inactivation of Notch signaling and the loss of the asymmetric and perpendicular cell divisions as well as the balance of self-renewal and differentiation in the mitotically dividing distal lung epithelial stem/progenitor cells (El-Hashash et al. 2011c).

Therefore, Eya1 protein phosphatase regulates the asymmetric mode of cell division in stem and progenitor cells of the distal lung epithelium and, thus, plays a major role in balancing their self-renewal with differentiation, which is required during lung development, repair, and regeneration. In several cases of congenital or developmental lung defects such as the congenital lung hypoplasia and bronchopulmonary dysplasia (BPD), which are probably caused by a stem/progenitor cell deficiency, the asymmetric mode of cell division is essential for normal lung morphogenesis as well as for the protection against the development of these severe congenital and developmental disorders. Balancing stem/progenitor cell self-renewal with differentiation is, therefore, essential for the regulation of the branching morphogenesis of epithelial tubes to produce sufficient surfaces for gas diffusion that is critical for human life. Both the defective differentiation, in which the asymmetric cell division is involved, and postnatal respiratory distress directly cause developmental defects (Warburton et al. 2008, 2010).

In summary, like other systems, several lines of evidence show that the asymmetric division is the common mode of division in stem and progenitor cells of the distal epithelium of developing lung. In addition, most stem and progenitor cells of the distal lung epithelium are polarized since they show an apical and asymmetric distribution of several polarity proteins such as mInsc, NuMA, and LGN as well as both Par-3 and Pr-6 proteins (El-Hashash et al. 2011c; El-Hashash and Warburton 2011; Figs. 5.3 and 5.4). Moreover, most of the dividing stem and progenitor cells in the distal lung epithelium are characterized by their mitotic spindle fibers that have a perpendicular alignment to the basement membrane, and the asymmetrical inheritance of the cell fate determinant Numb (El-Hashash et al. 2011c). Recent cadherin hole analyses have provided further evidences that the asymmetrical division is common in stem and progenitor cells of the distal lung epithelium (El-Hashash and Warburton 2012). Furthermore, Eya1 protein phosphatase functions to regulate the establishment of proper cell polarity and mitotic spindle orientation, maintain asymmetrical Numb inheritance, as well as *balance self-renewal* and *differentiation*. It is, therefore, a crucial regulator of the behavior of stem/progenitor cells in the distal lung epithelium (El-Hashash and Warburton 2011, 2012; El-Hashash et al. 2011c). The polarized localization of many polarity proteins and the perpendicular mitotic spindle alignment in distal lung epithelial stem and progenitor cells are in a strict harmony with the asymmetric cell division studies in other types of epithelial cells in both mammals and *Drosophila* (Cayouette and Raff 2002, 2003; Haydar et al. 2003; Noctor et al. 2004; Lechler and Fuchs 2005).

Chapter 6
Lung Stem Cell Plasticity

Abstract Cells could switch from one specific phenotype to another in response to specific regulatory signals from the surrounding environment. These changes may involve changes in cell fate, proliferation, shape, and adhesion and may be reversible or irreversible. More data suggest that epithelial cells are inherently plastic, and this plasticity is clearly exemplified by the process of epithelial to mesenchymal transition. Accumulated recent research on stem cell fate and behavior has focused on investigating their plasticity in adult tissues that undergo repair/regeneration and depend heavily on the lineage tracing technique. Stem cell plasticity is, therefore, the ability of one stem cell to produce other cell types that were considered outside their normal differentiation repertoire. Like other organs, lineage tracing has been largely used in the lung, to identify different distinct populations of epithelial stem/progenitor cells, which can display lineage plasticity after lung injury. In this chapter, we discuss evidences of the plasticity in lung stem/progenitor cells and our current thoughts on the various types of adult lung cell lineages and how they respond to lung injury. We also describe factors, signals, and molecular mechanisms that control the plasticity of lung-specific stem/progenitor cells.

Keywords Lung · Stem cells · Progenitor cells · Transdifferentiation · Cell plasticity

The cellular plasticity is currently an important theme in the development, function, and biology of many adult tissues and organs. This contrasts with many previous thoughts on the irreversibility of cell fate decisions in various types of developing organs and systems (Tata and Rajagopal 2017). For example, many studies have used the lineage tracing methods to define and characterize populations of lung-specific epithelial stem and progenitor cells that have a remarkable maintenance capability of their identity during various steady-state conditions. However, these studies can also show a characteristic lineage plasticity following lung injury. In this chapter, we will describe the current understandings of various cell lineages in the adult lung in mammals and discuss their responses to different types of lung injuries.

© Springer International Publishing AG, part of Springer Nature 2018
A. El-Hashash, *Lung Stem Cell Behavior*,
https://doi.org/10.1007/978-3-319-95279-6_6

6.1   Proof of Plasticity in the Lung Stem Cells

There are currently three general concerns in the complex field of stem cell plasticity in the lung. The first general concern is in the method of transdifferentiating cell identification in the lung. Thus, the transdifferentiating cells may be identified as such simply because their cell marker gene is one of the genes that are responsive to an injury. The second general concern is that the cell, which was incorrectly concluded to be formed by the transdifferentiation process in the lung, was not produced by converting one type of lung differentiated cell into another. Instead, it might be tagged because of the enhanced expression of one of the genes that are responsive to the injury (reviewed in Tata and Rajagopal 2017). There are several reasons for that to happen. For example, the antibody staining may be misleading because of the absence of the targeted protein during the setting of the active transgenic expression (Tata and Rajagopal 2017).

In addition, a remarkable study by Vaughan and co-workers has strongly suggested another important consideration for carrying out studies with a precise analysis of the lineage tracing, in which the specific labeling of cell types is difficult to be deleted from the injured tissue itself. This will help avoiding the incorrect tagging of cell types because of the issue of enhanced expression of one of the genes that are responsive to the injury that can be also avoided by demonstrating the plasticity of lung stem cells by using different driver types (Vaughan et al. 2015; Tata and Rajagopal 2017). Another general concern is that lung stem cells are probably so plastic that they can oscillate between different states that are closely related (reviewed in Tata and Rajagopal 2017).

6.2   Factors Controlling the Plasticity of Lung Stem Cells

Currently, one of the major questions in the field of lung stem cells is how the state of cellular maturity can affect the cell plasticity. This question has been raised because in several cases, the cell state plasticity can only be observed in specific cells that are undergoing differentiation, whereas in other cases the entirely mature differentiated cells have a dedifferentiation capacity into tissue-specific stem cell types. For example, several studies have shown that differentiating spermatogonial or gonialblasts cells can be dedifferentiated into stem cells, while mature differentiated cells are not capable to carry out the same process in *Drosophila* testis and ovary (Brawley and Matunis 2004; Kai and Spradling 2004). Remarkably, in other tissues or organs such as the murine trachea, fully mature secretory cells have a dedifferentiation capacity to stem cells after the loss of stem cells (Tata et al. 2013).

In the large airways of the lung, the regeneration of epithelial cells can occur through the differentiation of conventional basal stem cells. However, another type of cells, secretory cells, can also dedifferentiate and serve as stem cells, if the basal cells are lost. Indeed, secretory cells are capable to produce new basal cell populations. Interestingly, having at least 20% of the original basal cells intact can suppress the plasticity of secretory cells and their ability to produce basal cells in the large airways (Tata and Rajagopal 2017). Notch signaling activity plays a critical role in the plasticity of basal cells and in the maintenance of both identity and stability of secretory cells (Pardo-Saganta et al. 2015b). Multiple signals and molecular mechanisms are probably required to control the dedifferentiation process. However, it is not clear yet how Notch pathway and other signaling mechanisms control the dedifferentiation process (Tata and Rajagopal 2016, 2017).

Furthermore, another common thought in a wide range of tissues or systems is that acquiring a stem/progenitor cell fate and state occurs either permanent or transient. For instance, it is well-documented that the hair follicle stem and progenitor cells are capable to contribute to the epidermal wound regeneration, following a severe interfollicular epidermal injury. However, intensive lineage tracing analyses demonstrate the loss of persistence in these epidermal cell types over time (Ito et al. 2005). In contrast, other lineage tracing analyses show the persistence of basal stem cells that are dedifferentiated and their participation to the airway epithelium and repair of injured lung (Tata et al. 2013).

Chapter 7
Lung Stem Cells in Lung Repair and Regeneration

Abstract Intensive studies on lung development have helped to determine and characterize the lung-specific stem/progenitor cells and their regulatory molecular mechanisms. The adult lung consists of a wide range of different cell lineages, which are clearly quiescent in the absence of injury. The lung could remarkably respond quickly to different types of acute damage. This response is evidenced by the cell cycle, reentry of lung-specific stem cells, and their ability to differentiate to promote lung repair/regeneration. The process of lung repair and regeneration after acute injury, therefore, includes many of the stem and progenitor cell lineages. The accumulated research findings from lung developmental biology are currently widely used to determine the mechanisms that underlie lung repair/regeneration. This chapter describes our current knowledge of the roles of lung-specific stem cells in both lung repair and regeneration. It also describes how basic studies into lung developmental biology and regulatory molecular mechanisms are now being applied to lung repair/regeneration after injury.

Keywords Lung development · Stem and progenitor cells · Repair · Regeneration · Signaling pathways · Wnt · Notch · BMP · TGF

Rapid progress occurs in the repair and regeneration of different organ types. However, basic research and clinical practice in the lung regeneration remain crawling. The regeneration of the lung is a great challenge because of its complicated three-dimensional structure that contains more than 40 cell types. Distinct populations of intrapulmonary and extrapulmonary stem and progenitor cells can regenerate both epithelial and endothelial tissues in various parts of the respiratory tract. Recent lung stem cell researches and discoveries in humans have opened the door of hope again, which might put us on the path to repair our injured body parts, lungs being on demand (reviewed in Yang et al. 2015). This chapter describes lung stem cell capabilities and roles in lung repair and regeneration as well as the molecular mechanisms and developmental pathways that regulate these processes.

© Springer International Publishing AG, part of Springer Nature 2018　　　　61
A. El-Hashash, *Lung Stem Cell Behavior*,
https://doi.org/10.1007/978-3-319-95279-6_7

7.1   Stem Cell Capability for Lung Repair and Regeneration

The adult lung undergoes a slow turnover in humans. However, the lung characteristically repairs itself rapidly after injury, suggesting that there is a subpopulation of stem or progenitor cells that have a preserved differentiation potential in the adult lung. Several lung stem and progenitor cell types are capable to repair and/or regenerate the injured lung. These stem and progenitor cell types include the alveolar epithelial type II cells, club cells, and bronchoalveolar stem cells. In addition, other cell types such as the submucosal gland duct stem cells and neuroendocrine cells have a differentiation capacity to produce the basal, club, serous, or ciliated cell types, as well as the distal airway epithelium of the adult lung (Warburton et al. 2010; Green et al. 2013; Rawlins 2015; Stoltz et al. 2015; Chen and Fine 2016; El-Badrawy et al. 2016; Stabler and Morrisey 2017).

The repair and regeneration of the mammalian lung epithelium are crucial for restoring various normal lung functions after acute injury and are dependent on the involved spatial stem niches of stem cells. A variety of lung cell types and molecular mechanisms contribute to the tightly controlled processes of lung regeneration and repair after injury. Among important lung cell types, the alveolar type II *epithelial cells*, club cells, and bronchoalveolar stem cells are major stem and progenitor cell types that play predominant roles in the maintenance of epithelial turnover and repair after injuries (Warburton et al. 2010; Green et al. 2013; Kotton and Morrisey 2014; Rawlins 2015; Chen and Fine 2016; Stabler and Morrisey 2017).

The lung alveolar epithelium is characterized by two cell types: alveolar epithelial type I and II cells. The alveolar epithelial type I cell (AEC1) that comprises almost 95% of the alveolar surface area is a thin but architecturally complex cell type, which is specialized for gas exchange (Weibel 2015). The AEC1 cells express several distinct markers, including aquaporin 5 and podoplanin. A small subset of AEC1 cells that is characterized by the expression of HOPX can proliferate and give rise to alveolar epithelial type II (AEC2) cells after partial pneumonectomy (Jain et al. 2015). However, whether AEC1 cells can self-renew or serve as progenitors in other types of lung injury remains controversial.

The AEC2 cell is a cuboidal cell that is situated at the corners of the alveoli. They are characterized by their production and packaging of surfactant proteins into lamellar bodies for secretion. A variety of cell-specific markers for AEC2s have been identified, including LYZ2 and SFTPC (Warburton et al. 2010). By using a mouse expressing an inducible *Sftpc*-driven Cre recombinase, adult AEC2 cells were shown to serve as progenitors for both AEC1s and AEC2s during homeostasis (Barkauskas et al. 2013). In addition, cell turnover is remarkably accelerated and characterized by clonal expansion of a subset(s) of AEC2s that is followed by differentiation into AEC1s, during alveolar repair after AEC2 ablation, hyperoxia, or bleomycin administration (Desai et al. 2014). Several signaling molecules may regulate the self-renewal of AEC2 after lung injury such as oncogene *KRAS* and epidermal growth factor receptor (Desai et al. 2014). Notably, the pulmonary capillary endothelial-derived MMP14 was implicated in the alveolarization regeneration after pneumonectomy by enhancing the bioavailability of epidermal growth factor receptor ligands (Ding et al. 2011).

7.2 Regulatory Mechanisms and Developmental Pathways

During lung repair and regeneration, the signaling pathways that are used by lung-specific stem and progenitor cells are often the same as those used for building up the lung during its developmental and morphogenetic stages (Shi et al. 2009; Whitsett et al. 2011; Akram et al. 2016; Bertoncello 2016; Herriges and Morrisey 2014; Stabler and Morrisey 2017). Notably, the functional roles of many developmentally conserved regulatory factors, molecular mechanisms, and signaling pathways in the repair and regeneration processes are not limited to the lung epithelium but are also important for various mesenchymal lineages (Stabler and Morrisey 2017). Thus, the developmental Wnt signaling pathway is fundamental for the lung development, and several studies revealed that the Wnt pathway has crucial functional roles in the expansion of the bronchoalveolar stem cell lineage in the more distal lung airways (Zhang et al. 2008). The function of Wnt signaling in the bronchoalveolar stem cell expansion is regulated by Gata6, which is an essential transcription factor for lung formation and development (Zhang et al. 2008).

BMP4 growth factor is another important regulator of the lung repair and regeneration since it regulates both the bronchoalveolar stem cell differentiation and the basal cell behavior in the lung. BMP4 stimulates the protein expression level of Thrombospondin-1 (TSP-1), which is an extracellular matrix protein, in the lung endothelium (Lee et al. 2014). TSP-1, in turn, controls the differentiation of bronchoalveolar stem cells (*BASCs*; Lee et al. 2014).

The basal stem/progenitor cells and both ciliated and secretory luminal cells represent the pseudostratified epithelium in the lung. A recent study identified signals that regulate the basal cell behavior by screening for factors that can change basal cell differentiation and self-renewal/proliferation in a clonal tracheosphere organoid assay (Tadokoro et al. 2016). In this study, disruption of the activity of BMP signaling leads to a stimulation of cell proliferation, whereas the cell treatment with exogenous BMP4 results in a suppression of both cell proliferation and differentiation (Tadokoro et al. 2016). Interestingly, changes in different components/elements of the BMP signaling pathway were detected in the murine trachea during the regeneration of the epithelium from basal cells after an in vivo injury (Tadokoro et al. 2016).

BMP signaling pathway is, therefore, normally a negative regulator of the proliferation of these cells. However, this effect can be transiently released during the repair process when the endogenous antagonists of BMP are upregulated (Tadokoro et al. 2016).

The Notch signaling pathway controls the development of the airway epithelium and can regulate both the activation and differentiation of Trp63+ basal cells into multi-ciliated and secretory epithelial cells in both the large airways and trachea

(Guseh et al. 2009; Rock et al. 2011; Tsao et al. 2011; Mori et al. 2015). Various members of the Notch family have characteristically distinct effects in the lung. Notch 3, for example, functions to inhibit the overexpansion of basal cells (Pardo-Saganta et al. 2015), while Notch 2 activity plays an essential role in alveologenesis (Tsao et al. 2016).

The adult lung represents a highly quiescent tissue/organ with a remarkably low level of cellular turnover. The cellular quiescence of the murine airway epithelium and mesenchyme is an active process that is mediated by sonic hedgehog signaling (Peng et al. 2015). Suppression of hedgehog signaling can lead to the increase of the epithelial cell proliferation at homeostasis and after lung injury (Peng et al. 2015). Although it is not yet clear whether the hedgehog signaling pathway functions similarly in the human lung, these findings by Peng et al. (2015) may be relevant to the pathogenesis of lung diseases that are characterized by defected repair and regeneration.

Transcription factors are also important for the regeneration of the adult lung. In the lung, the pseudostratified airway epithelium contains both polarized secretory and ciliated cells. Basal stem and progenitor cells are critical for the maintenance of both polarized secretory and ciliated cells. However, little is known about the molecular mechanisms, factors, and signaling pathways that coordinate the lineage choice and differentiation of these distinct cell types with the lung epithelial morphogenesis. These mechanisms and factors were investigated by a remarkable study from Brigid Hogan's laboratory that also addressed important questions about the role of transcription factors in regenerating the mucociliary epithelial cells from basal progenitor cells (Gao et al. 2015). They found that the grainyhead-like 2 transcription factor is a critical positive regulator of ciliated cell differentiation and functions to promote both the ciliogenesis and barrier function in the lung by regulating Znf750 transcription factor and some glycoproteins (Gao et al. 2015).

The unique plasticity of some lung airway epithelial cell types is also important for lung repair/regeneration. For example, the secretory cells in the lung are capable of dedifferentiation backward into the basal cells (Tata et al. 2013). This dedifferentiation process is required for replenishing the populations of basal cells after lung injury (Tata et al. 2013). Despite the recent progress in the understanding of lung epithelial cell plasticity, the exact functions of lung plasticity in the repair and regeneration of the lung, as well as the regulatory molecular mechanisms and signaling pathways that control these processes, still need further studies (Stabler and Morrisey 2017).

Furthermore, the epithelial Na(+) channel (ENaC), which is a crucial signaling pathway for both trans-apical salt and fluid transport and expressed by lung epithelial stem and progenitor cells, is involved in the development, repair, and regeneration of the lung epithelium (Liu et al. 2016). The epithelial Na(+) channels (ENaC) are expressed in the alveolar type II epithelial and club cells, and its expression pattern and activity are development- and differentiation-dependent (Liu et al. 2016). In addition, the ENaC activity is probably required for the differentiation and migration of both circulating and local stem cells that can support the stem cell functions in the reepithelialization of the injured lung (Liu et al. 2016).

Finally, the importance of different signaling pathways is not limited to their roles in the lung development, repair, and regeneration but extends to both the aging and some severe pulmonary diseases that currently represent a significant and increasing health burden. Aging is widely thought to be driven, at least partially, by an aberrant activity of several types of developmental signaling pathways, including the Wnt pathway, which might drive several pathological changes. This process has been termed as the developmental drift or antagonistic pleiotropy (reviewed by Lehmann et al. 2016). In addition, several studies have linked the age-related pathologies to changes in the Wnt activity and, therefore, focused on both the determination of the Wnt signal regulation during aging and the Wnt effects on age-related pathologies in some organs (Lehmann et al. 2016). Furthermore, the Wnt pathway is critical for the development of several lung diseases besides its functions in the lung development and regeneration. The altered Wnt signaling activity contributes to the pathogenesis of severe pulmonary diseases, including the chronic obstructive pulmonary disease and the pulmonary fibrosis (IPF; Lehmann et al. 2016).

Chapter 8
Stem Cell-Based Organoid Models in Lung Development and Diseases

Abstract The recent application of co-culture organoid systems in different organs has successfully helped in the in vitro cultivation of stem cell populations that were previously inaccessible. These co-culture organoid systems have provided a novel method for investigating the cellular and molecular mechanisms controlling the development, interaction, and function of these cell types. In the lung, organoid cultures have been recently used for cell-cell interaction studies. These cultures rely on the interactions between the lung stem cells and a putative niche cell that is important for their behavior, differentiation, and growth. The organoid systems have been used in the study of airway basal cells, but the applications of organoid systems for the study of other lung regions or cell types are still in its infancy. This chapter describes our current knowledge of the stem cell-based organoid models in lung development and diseases. It also describes recent advances in the embryonic lung-derived organoids, the adult lung-derived organoids, and organoids from iPSC-derived lung epithelial cells.

Keywords Organoids · Stem and progenitor cells · Co-culture · Mouse lung · Human lung · iPSCs

8.1   Introduction on Early Organoid Studies

The rapid progress of 3D culture technology has clearly allowed different types of embryonic and adult stem cells in mammals to display their characteristic properties of self-organization, which result in the formation of organoids that usually reflect many characteristic functional and structural features of a wide range of organs such as the lung, kidney, and others. Therefore, the organoid technology has high hopes and applications in the modelling of human organogenesis and various human diseases "in a dish" (Clevers 2016). Another advantage of this technology is that organoids that are derived from patients can be used for predicting drug response in a personalized manner, and, therefore, the organoid technology provides effective tools for regenerative medicine and gene therapy as well as both precision and personalized medicine (reviewed in Clevers 2016; Barkauskas et al. 2017).

© Springer International Publishing AG, part of Springer Nature 2018   67
A. El-Hashash, *Lung Stem Cell Behavior*,
https://doi.org/10.1007/978-3-319-95279-6_8

Organoids or "mini-organs" are three-dimensional (3D) cellular structures that are formed from stem cells and have many types of organ-specific cells. The cell-sorting abilities and lineage commitment that are spatially restricted enable these cell types to self-organize in a specific manner that is similar to their native organ. They also show degrees of organ functionality (Nadkarni et al. 2016; Clevers 2016; Nikolić and Rawlins 2017; Barkauskas et al. 2017, for review). The 3D culture of organoids can prevent the cells from being transformed and preserve native DNA integrity, and, therefore, they are useful tools for modelling both human organ development and disease as well as for screening drugs in culture (Huch et al. 2015; Nikolić and Rawlins 2017). The growth of the organoid requires several biological processes, including the initiation of stem cell population to self-renew, leading to the increase of organoids' size and cell differentiation (Nikolić and Rawlins 2017; Barkauskas et al. 2017).

Early successful organoids were formed from murine small intestine by using single Lgr5+ positive stem cells. These organoids were completely epithelial indicating that tissue-specific organoids could be established without other non-epithelial cell types (Sato et al. 2009). In these early studies, the organoid was structured as units for the crypt-villus that have a similar hierarchy of stem cells to in vivo. This indicates that the interactions between epithelial cells are adequate for generating units of the crypt-villus (Sato et al. 2009). Moreover, Sato and colleagues used these intestinal organoids to investigate cell-cell interactions and found that Paneth cells that are scattered through Lgr5+ stem cells provide important niche signals (Sato et al. 2011a).

Intensive research has led to a successful culture of organoids from several endoderm-derived organs such as the adult mouse colon, stomach, pancreas, liver, and prostate, as well as the adult human colon, intestine, and prostate (Barker et al. 2010; Sato et al. 2011a, b; Huch et al. 2013a, b; Karthaus et al. 2014; Yin et al. 2014). Successful organoid cultures were also reported to mouse embryonic pancreas (Greggio et al. 2014). Interestingly, the same tissue culture medium has been reported to support both the self-renewal and differentiation of stem cells in most of these research studies. For example, in the studies of Sato and colleagues, intestinal stem cells were shown to self-organize into organoids and differentiate in the same culture medium (Sato et al. 2009, 2011a). The organoids of other organs such as the adult liver and pancreas can also be expanded. However, these organoids cannot yet easily differentiate (Huch et al. 2015; Broutier et al. 2016). In addition, several studies on murine embryonic pancreas progenitor cells showed that these cells could expand in a self-renewing medium and then mature when growing in a differentiation medium (Greggio et al. 2013, 2014, 2015). In addition to the composition of the culture medium, the cell-cell interaction within organoid cultures is another important factor for the organoid differentiation in culture. This phenomenon is now known as the "community effect," which indicates that the ability of cells to differentiate is increased by neighboring cells differentiating in the same way simultaneously (Gurdon 1988).

Like other endoderm-derived organs, several studies have successfully grown lung organoids from the embryonic lungs (Mondrinos et al. 2007, 2014; Zhang et al.

2014; Sucre et al. 2016; Wilkinson et al. 2016), mouse adult lungs (Rock et al. 2009, 2011; Barkauskas et al. 2013; Lee et al. 2014), human adult lungs (Rock et al. 2009; Barkauskas et al. 2013), and lung progenitor cells that are derived from human-induced pluripotent stem cells (Gotoh et al. 2014; Dye et al. 2015; Konishi et al. 2016).

8.2   Embryonic Lung-Derived Organoids

The development of the murine lung starts after the formation of the primary germ layers. In the mouse, the lung starts growing as two primary small buds, which arise from the ventral floor of the primitive foregut at around E9.5 of pregnancy (Hogan et al. 2014). The expression of Nkx2.1 transcription factor can mark the initial development of the respiratory lineage in the ventral anterior foregut endoderm starting from E8.25 of mouse embryonic development (Serls et al. 2005; Morrisey and Hogan 2010; Herriges and Morrisey 2014). In both the human and mouse embryos, the lung develops as a blind-ended tube that undergoes intensive and complex rounds of outgrowth and branching morphogenesis to form a tree-like lung organ the performs respiration. These processes are controlled by a wide range of transcription factors and signaling pathways (Warburton et al. 2010; El-Hashash 2013; Rankin and Zorn 2014; Ibrahim and El-Hashash 2015; Schittny 2017; Chen and Zosky 2017). The cell populations that represent multipotent stem or progenitor cells, which form alveolar and bronchiolar cells, are localized in the developing distal branching tips of the embryonic mouse lung (Rawlins et al. 2009a, b; Desai et al. 2014; Alanis et al. 2014). Several studies have grown whole embryonic mouse lungs (E12.5 of development) in culture on an air-liquid interface in 2D. Under these conditions, these embryonic lungs can differentiate. They, therefore, represent ultimate or basic mini-organs as they structurally have all the cell populations that are required for a proper development of the lung. However, these explant cultures are not suitable for a lot of experimental strategies because of their ability to grow in a 2D but not in 3D plane, complex cellular structure, and size (Jaskoll et al. 1988; Seth et al. 1993; del Moral and Warburton 2010).

During embryonic development, both the thyroid and the lung arise from Nkx2-1+ progenitor cells in the ventral foregut endoderm. A remarkable and leading research study by Longmire and co-workers have directly differentiated primordial progenitor cells of the lung and thyroid from embryonic stem cells based on using the activin-induced definitive endoderm and treatment with TGF-beta and BMP inhibitors, followed by the stimulation of BMP/FGF signaling activity (Longmire et al. 2012). This successful protocol generates a relatively pure progenitor cell population, which could recapitulate key early developmental steps of lung or thyroid development (Longmire et al. 2012).

Mixed populations of cells that are obtained from the dissociation of whole E17.5 murine lungs can be grown in 3D on different matrix types and are capable of self-organization into spheroids that contain many branched epithelial structures surrounded by mesenchymal cells. Thus, in these cultures the whole fetal lungs are minced before their growth as spheroids, and, therefore, these cultures have certain limitations because their starting population is not well-defined (Nikolić and Rawlins 2017). Although these cultures can grow well and show several evidences of both bronchiolar and alveolar differentiation, their self-renewal capacity and cellular phenotyping still require more extensive investigations (Mondrinos et al. 2007, 2014; Zhang et al. 2014). However, these spheroids are currently useful tools for the study of lung morphogenesis and development in culture. For example, they can be used to determine the effects of several growth factors, including VEGF-A and FGFs, or for direct lung cell differentiation to alveolar cell types (Mondrinos et al. 2007, 2014; Zhang et al. 2014).

The successful formation and growth of mesenchymal organoids that consist of fibroblast cells isolated from human fetal lung (18–20 week of gestation) and grown on collagen-coated alginate beads in 3D units have been reported (Sucre et al. 2016). To mimic the oxygen tension that is experienced by many premature babies for disease modelling, the protocol of these studies involves the exposure of the mesenchymal organoids to various concentrations of oxygen (Sucre et al. 2016). This system can potentially incorporate other types of lung cells such as airway epithelial and endothelial cells (Wilkinson et al. 2016). Further studies are still needed to test whether these approaches and models can potentially and accurately recapitulate cell interactions and disease phenotypes.

8.3   The Adult Lung-Derived Organoids

Organoid cultures have been used to investigate the identity of lung stem cells and dissect molecular and cellular mechanisms and interactions for the adult lung. For example, single basal cells were isolated from the adult mouse trachea and grown into organoids that are called tracheospheres. In these tracheospheres, the basal cells can both expand and then differentiate, after changing the medium composition (Rock et al. 2009). Interestingly, these well-studied cultures could be passaged two or more times, which further confirms that basal cells are adult stem cells with a self-renewal ability (Rock et al. 2009). These cultures have subsequently been used successfully to investigate the functional role of several signaling pathways such as BMP, Notch, and IL6 signaling and certain types of transcription factors in basal cells of the adult lung (Rock et al. 2011; Tadokoro et al. 2014, 2016; Gao et al. 2015).

Moreover, an in vivo strategy of lineage tracing that is combined with a derivation of tracheospheres has been used to investigate the interactions between basal and club cells (Tata et al. 2013). This strategy has uncovered the ability of basal cells to suppress the club cell dedifferentiation in culture (Tata et al. 2013). For the initial derivation of tracheospheres, a still undefined culture medium was optimized and used for basal cell expansion and differentiation in 2D conditions at air-liquid interface (You et al. 2002). Several recent studies have tried to improve and optimize the growth conditions for basal cells in 2D that can also facilitate and improve research experiments on organoids. For instance, basal cells were co-cultured with fibroblast cell lines, which were initially used for the epidermal expansion in culture (Hynds et al. 2016). Another approach was to suppress SMAD signaling activity that clearly led to the basal cell expansion from several organs, including the murine and human trachea (Mou et al. 2016). The last two studies have raised important questions about the signaling interactions of the lung airway and basal cells in their niche in vivo and how this could be answered, at least partially, by using organoids *that were* used for *experiments* such as *co-culture* (Hynds et al. 2016; Mou et al. 2016).

In mouse, the smaller airways contain both ciliated cells and secretory club. Several studies have used the co-culture strategy of club cells with underlying stromal cell populations to grow these club cells as spheroids. Isolated club cells must, therefore, be co-cultured with fibroblast cell populations to grow as spheres (Teisanu et al. 2011). Similar co-culture techniques have been used successfully by subsequent research studies that also aimed to explore the functional roles of signaling pathways such as FGFR signaling between the epithelial and mesenchymal cell compartments in the co-cultures. These studies have indeed demonstrated the importance and potential applications of organoid-based co-culture systems in the analysis of signaling between different cell types in culture (Hegab et al. 2015). However, it is still difficult to systemically analyze the cross talk between cells of these co-cultures since the identity of the mesenchymal cell component is still unclear. In contrast, some studies have shown that co-culture (as organoids) with pulmonary endothelial cells is sufficient to facilitate the development and differentiation capacities of the putative BASC stem cells in vitro, which are epithelial cells co-expressed genetic markers of both alveolar and bronchiolar cell lineages and located in the terminal bronchioles (Lee et al. 2014). Similarly, this co-culture system has been used successfully to dissect signaling pathways such as BMP4 that regulate epithelial-endothelial cell interactions that function to support the differentiation of alveolar lineages. In addition, these studies suggest a strong relationship and/or interaction between different cell types of alveolar endothelial, epithelial, and hematopoietic cell compartments in vivo during the repair and regeneration of the alveolar tissue that could be determined and characterized using organoid-based systems (Ding et al. 2011; Rafii et al. 2015; Cao et al. 2016).

Furthermore, several in vivo and in vitro studies have provided evidences that the alveolar epithelial type II cells (AC2) are the major contributing stem cells to the alveolar epithelium, at the steady state (Barkauskas et al. 2013; Desai et al. 2014). Moreover, alveolar epithelial organoids, known as alveolospheres, were successfully generated by the co-culture of isolated type II alveolar epithelial cells with a PDGFRA+ fibroblast cell population that may correspond to the alveolar lipofibroblast cells (Barkauskas et al. 2013). Using this co-culture system for organoids, AC2 cells were shown to have more sensitivity to a reduction in the telomere length than the stromal cell populations that support their growth in culture (Alder et al. 2015). This study has also shown that the failure of epithelial stem cells is the primary defect in various types of lung diseases that are caused by telomere syndromes

(Alder et al. 2015). More analyses of alveolar organoids that are based on co-culture systems are still needed to further uncover the factors and signaling requirements for the growth and differentiation of alveolar epithelial cells.

The steady-state characteristics of distal lung stem cell populations are not completely defined. However, these cells play an important role in a wide range of murine alveolar repairs after injury, including those that are caused by influenza infection, and were expanded in 2D cultures. It is widely agreed that these cells are probably Sox2-positive and can upregulate the markers of basal cells, including p63 and keratin5 (KRT5) after injury. These cells may also have a multilineage cell differentiation in organoid cultures in vitro (Kumar et al. 2011; Zuo et al. 2014; Vaughan et al. 2015; Ray et al. 2016). In addition, a spheroid co-culture system for stem cell populations of the distal lung has been reported recently (Quantius et al. 2016). This study has not provided strong evidences of cell self-renewal or differentiation. However, it will be useful for future mechanistic studies on lung stem cells.

Human lung diseases could not be completely phenocopied using murine models of lung diseases. The generation of mature lung cells specific for human lung diseases is, therefore, highly significant for disease modelling. Notably, the methods required for human airway basal stem cells to grow in culture at air-liquid interface that aim to recapitulate the organization of the airway epithelium are currently both well established and adapted to include supporting cells such as vascular endothelial and fibroblast cells (Curradi et al. 2012; Fulcher and Randell 2013; Hill et al. 2016). Additionally, several studies have shown evidences that human airway basal cells can also grow and differentiate as organoids and allow efficient genetic manipulations in organoid culture (Rock et al. 2009; Gao et al. 2015), and more research is currently underway to further optimize these techniques in different research laboratories. In contrast, the growth of human AC2 cells in culture as organoids has achieved less success, compared to that for murine AC2 cells (Barkauskas et al. 2013). In addition, the self-renewal or differentiation of human AC2 cells is still unsuccessful in 3D cultures.

8.4   Lung Epithelial Cells Derived from iPSCs as a Source of Organoids

Patient-specific iPSCs can be differentiated into any relevant cell type. This differentiation capacity has many important applications, including those for cell-based therapies, modelling of human diseases, and screening of drugs. Early studies have shown that adult somatic cell reprogramming can lead to the formation of iPSCs, which was performed by the introduction of some pluripotency transcription factors into these somatic cells (Takahashi and Yamanaka 2006). Different research laboratories have successfully grown and differentiated iPSCs into lung progenitor cells (Mou et al. 2012; Longmire et al. 2012) as well as into well-differentiated bronchiolar and alveolar epithelial cells (Wong et al. 2012, 2015; Ghaedi et al. 2013; Firth et al. 2014; Huang et al. 2014, 2015). These research laboratories have used methods that recapitulate normal murine lung development and used protocols that used directed differentiation of iPSCs as monolayers in 2D culture systems. Thus, they used research methods that include differentiation into definitive endodermal structures that eventually lead to the formation of both anterior and ventral foregut endodermal derivatives. Subsequently, they used a stage-specific combination of certain growth factor types to induce the differentiation of ventral anterior foregut endoderm (VAFE) Nkx2.1+ lung progenitor cells into either alveolar or bronchiolar cell fates (Longmire et al. 2012; Mou et al. 2012; Wong et al. 2012, 2015; Ghaedi et al. 2013; Firth et al. 2014; Huang et al. 2014, 2015). The maturity of iPS-derived differentiated cells has been recently enhanced by 3D differentiation as organoids (Dye et al. 2015; Fox et al. 2015). This approach has been effective to some extent for the differentiation of lung alveolar (Gotoh et al. 2014) and bronchiolar cells, and there were some evidences of a successful differentiation into ciliated cells with beating cilia (Konishi et al. 2016).

The generation of lung organoids from induced pluripotent stem cells has been successful using a protocol that involves an air-liquid interphase culture at its last stage (Wong et al. 2012). The protocol designed in Wong et al. (2012) study was used for CFTR mutant iPS cells as a proof of concept for modelling cystic fibrosis. In another study Snoeck and co-workers have designed an improved four-stage, 50-day protocol for more effective generation of airway-specific epithelial cells from human pluripotent stem cells (Huang et al. 2014). In this protocol, the definitive endoderm was induced by activin A, which has a wide range of biological activities in different cell types. This was followed by the induction of the anterior foregut endoderm formation through a sequential inhibition of the activities of BMP, TGF-β, and Wnt signaling. Subsequently, the cells were ventralized by Wnt, BMP, FGF, and RA signaling to obtain both lung- and airway-specific progenitor cells. During the last step of this protocol, different types of epithelial cell types (Clara, ciliated, basal, goblet, and alveolar epithelial cells types I and II) are matured using singling molecules such as FGF, Wnt, glucocorticoids, and c-AMP (Huang et al. 2014). Other researchers (Dye et al. 2015) have applied an activin treatment to human PSCs and then subsequently added TGF-β/BMP inhibitors, FGF4, and Wnt activators to instruct the cells toward an anterior foregut fate. When they activated the Hedgehog pathway simultaneously, the organoids were ventrally specified toward a lung fate (Dye et al. 2015). Interestingly, when Spence and co-workers embedded this organoid in Matrigel with a prolonged exposure to Fgf10, mature lung organoids arose (Dye et al. 2015). These organoid cultures resembled proximal airways, containing Clara cells, basal cells, and ciliated cells, and could be maintained for several months. Notably, this study has shown that smooth muscle actin (SMA)-positive mesenchymal cells often surround the endodermal airway tissues, and the early markers of the distal (alveolar) airways are expressed early in culture but are lost later (Dye et al. 2015).

Another pioneer study aimed to derive alveolar-specific organoids from iPSC-produced lung-specific progenitor cells in human (Gotoh et al. 2014). In this first published attempt, VAFE iPSC-derived lung progenitor cells were specifically isolated using the cell surface marker carboxypeptidase M (CPM). These progenitors were subsequently growing in culture by seeding into Matrigel with both a combination of growth factors that is specific to alveolar cells and human fetal lung fibroblast cells, which were obtained at 17.5 weeks of gestation (Gotoh et al. 2014). These organoids were NKX2.1+ and contained a small number of pro-SFTPC+ (SPC) AC2 cells and some Aquaporin5+ AC1 cells (Gotoh et al. 2014). In addition, this study has also shown that CPM+ iPSC-derived lung progenitor cells that grew in culture in the absence of human fetal lung fibroblast cells cannot differentiate and form SPC+ cells, which suggests that the fibroblast cells produce important signaling molecule(s) for the identity of alveolar cells. Moreover, in this study, the efficiency of production of AC2-like cells was low, and these cells were probably immature (Gotoh et al. 2014). Further studies are needed to improve this encouraging co-culture strategy, for example, by optimizing culture conditions and improving characterization of the fibroblast cell type(s).

A similar approach was used in another study to produce bronchiolar lung organoids based on the isolation of CPM+ iPSC-derived lung progenitor cells that was followed by developing an organoid cell culture (Konishi et al. 2016). This study has described cell differentiation into ciliated, secretory, and basal as well as neuroendocrine cells (Konishi et al. 2016). Remarkably, there was no clear ciliated cell differentiation in these 2D cultures, suggesting that 3D organoid culture rather than 2D culture enables more successful differentiation into functional ciliated cells (Konishi et al. 2016). However, the reported ciliary beating in this research was not fully synchronized as required for both mucociliary clearance and unidirectional flow, suggesting either functional immaturity or absence of some directional cues. Moreover, these organoids did not show any evidence for mesenchymal differentiation, and consisted of epithelial cells only.

Several studies have described the development of lung organoids that contain both alveolar and bronchiolar regions as well as associated mesenchymal cell derived from pluripotent stem cells. For example, a study has recently attempted to differentiate human pluripotent cells into the foregut (VAFE) spheroids (Dye et al. 2015). These VAFE spheroids were differentiated later into lung-specific organoids that have both alveolar and bronchiolar structures and grew in culture for over 3 months (Dye et al. 2015). Interestingly, these bronchiolar structures were surrounded, at least partially, by mesenchymal-derived actin-expressing smooth muscle cells, and included different cell types such as basal, club, and ciliated cells (Dye et al. 2015). In these organoid cultures, the mesenchymal *proximity* to the epithelial cells will likely provide a useful platform to investigate the interactions between the epithelial and mesenchymal cells in the embryonic lung during development. However, further research is still needed to determine the significance of the mesenchymal cells in these organoid cultures. In addition, lung organoids that are derived from human iPSCs are probably like human fetal lung tissues, as shown by transcriptome analyses, despite their 3-month long-term culture (Dye et al. 2015). Moreover, genetic analyses of the alveolar structures in these organoids have demonstrated that they co-express SPC and SOX9 or HOPX and SOX9 that is consistent with bipotent alveolar progenitor cells that exist in early developing murine lungs, rather than differentiated alveolar cells (Treutlein et al. 2014; Desai et al. 2014). Furthermore, growing lung organoids that were derived from human iPSCs on poly(lactide-co-glycolide) (PLG) scaffolds, followed by their xenotransplantation into kidney capsules of immunocompromised mice, can lead to the increase of organization of bronchiolar mesenchymal and epithelial cells compared to other organoids that were grown in culture (reviewed by Nikolić and Rawlins 2017). Another remarkable finding is that in vivo transplanted organoids can lead to the improvement of cellular differentiation of mesenchymal and secretory lineages as well as the associated vasculature (Dye et al. 2016).

All together, these studies have demonstrated that both alveolar and bronchiolar iPSC-derived organoids have the capacity to grow in culture to a degree of maturity that is like fetal lung tissues. Among these studies, studies that produce organoids consisting of both mesenchymal and epithelial lineages were the most successful, which suggests a critical role for the epithelial-mesenchymal cell interactions in cell differentiation. Several studies on the differentiation of many human organs from iPSCs such as the human kidney and intestine have shown similar results (Watson et al. 2014; McCracken et al. 2014; Takasato et al. 2016). In addition, advanced research has shown that cardiomyocytes that were derived from human iPSCs exhibit some degrees of cellular immaturity (so-called fetal phenotypes). This occurs despite the finding that 3D culture can increase maturity levels compared to well-established 2D cultures (Passier et al. 2016; Nikolić and Rawlins 2017). Several research approaches have been suggested for improving the maturity of cardiomyocytes such as high oxygen levels and prolonged culture period, as well as different combinations of either growth factors or co-cultures (Passier et al. 2016). Similar research approaches could be also used for lung-specific organoids, particularly with the promising recent findings that transplanted lung-specific organoids exhibit advanced signs of maturity in vivo (Dye et al. 2016). Further studies on the maturation and validation of these organoid lung cultures by comparing with human adult tissue are still needed. The findings of these studies will be critical in the determination of whether these lung-specific organoids are appropriate for modelling lung development and diseases, as well as regenerative attempts for the human lung.

Concluding Remarks and Future Perspectives

This book describes key aspects of stem and progenitor cell fate and behavior in the lung, comparable to other tissues and systems, in which stem cell fate and behavior are well studied.

Recent research has focused on the asymmetrical mode of cell division in stem cells of different species and provided insights into the molecular mechanisms and signals that maintain stem cells and give rise to cellular diversity. Important intrinsic factors and extrinsic signals have provided paradigms for how the asymmetric cell division is specified. Several reports have shown that similar mechanisms are found in both invertebrates and vertebrates. However, in mammalian stem cell systems, more improvements are still required in the areas of isolating pure stem cell populations, stem cell expansion in vitro, stem cell characterization in vivo, and the real-time imaging. These improvements would help identifying the mechanisms that regulate the asymmetric mode of cell division. In addition, the balance between the symmetric and asymmetric cell division produces the proper stem cell number that varies during organ development and regeneration. More studies are needed to identify the molecular mechanisms and dissect the signal pathways that control this balance that will help in advancing tissue repair and regeneration using stem cells.

Recent studies in our laboratory have characterized the asymmetrical division in lung epithelial stem cells and led to the identification of some regulatory mechanisms of this mode of division in the lung. Most polarized stem/progenitor cells in the lung distal epithelium divide asymmetrically and show perpendicular rather than parallel divisions, like many other tissues/systems (Lechler and Fuchs 2005; Yamashita et al. 2010; El-Hashash and Warburton 2011; El-Hashash and Warburton, 2012). The asymmetrical segregation and inheritance of Numb is probably the mechanism that controls the asymmetrical division and cell fate determination in lung stem/progenitor cells (El-Hashash et al. 2011c; El-Hashash and Warburton 2011), like stem cells of other tissue/organs (Morrison and Kimble 2006). The similarity of the asymmetrical stem cell division in the lung with other organs can help in finding novel solutions for the restoration of normal lung development and morphogenesis.

© Springer International Publishing AG, part of Springer Nature 2018
A. El-Hashash, *Lung Stem Cell Behavior*,
https://doi.org/10.1007/978-3-319-95279-6

The extrinsic and intrinsic factors and molecular mechanisms that direct the asymmetric divisions are well studied in *Drosophila,* as an invertebrate model. In mammals, the in vivo characterizations of stem cells, the mechanisms that regulate the asymmetric stem cell division and many other areas, *still require* further *intensive investigation*. For example, in mammalian lungs, determination of the molecular mechanisms that control the asymmetric stem cell division is crucial for lung development and regeneration. In addition, identification of factors and signals that can prevent or retard stem/progenitor cells from oscillating between the symmetrical and asymmetrical modes of division still require further investigations since the switch between these two modes of cell division is critical for lung repair. Furthermore, rebellions to normal control signals that direct the growth and repair are a hallmark of cancer cells. Elucidating these signaling mechanisms, therefore, allows for the ability to find new strategies to inhibit the initiation of cancer and even develop anticancer therapies in the lung and other organs.

The self-renewal of lung-specific stem/progenitor cells is fundamentally essential for the lung development. In addition, the differentiation of these cells produces a sufficient large alveolar gas diffusion surface, which is essential for different lung functions that maintain normal postnatal life. Abnormalities in these important biological processes may lead to a postnatal respiratory distress (Warburton et al. 2008, 2010). Furthermore, balancing stem cell proliferation/self-renewal with differentiation is critical for proper formation and regeneration of tissues in different organs, including the lung. The lack of this balance in the lung may result in several diseases such as the bronchopulmonary dysplasia (BPD) and pulmonary hypoplasia, wherein a remarkable deficiency of stem/progenitor cells may occur. These diseases or defects are closely related to lung injuries and/or *premature birth* and common health problems in newborn infants. They, therefore, cause a significant increase of death during infancy in humans. More intensive research is still needed to identify the regulatory mechanisms of balancing stem cell proliferation/self-renewal with differentiation since this will enable the development of new techniques to control the ability and application of stem in the repair and regeneration of the lung.

An intensive description of newly identified cell types using more advanced research methodologies, including the computational lineage reconstruction and the single-cell sequencing, is required for future research on the lung stem cell plasticity in human and murine models. In addition, new lineage tracing analyses with novel markers are also required to both verify and confirm these lineage predictions as well as to support several current hypotheses on the lung stem cell plasticity. Moreover, advances in the clonal epigenetic profiling techniques will more likely determine the ability of different types of cells to gain other cell states. Furthermore, recent, more advanced and powerful new tools in epigenetics and other related research areas will help to explore more about the transitions of cell state in different tissues and organs, including the lung.

In this book, we described several recent research methods can be applied for deriving 3D organoids from various populations of epithelial cells, which include AEC2 cells, basal cells, and secretory club cells in the adult lung. These organoid culture systems can be also derived from human pluripotent stem cells (hPSCs), and this book has described their importance in the understanding of various lung biological processes. The potential and future applications of murine- or human-based organoids in translational and/or basic research such as drug screening and modelling various types of human diseases have been also described in this book. Yet, more intensive research is still needed to overcome significant limitations and improve the use of organoid cultures in both modelling human diseases and drug screening as detailed below.

More studies are still needed to define various types of organoid culture media since the currently used culture media for organoid cultures are not chemically well-defined to date and contain many supplements. In addition, future research should also focus on rigorously testing the effect of other parameters affecting organoid cultures such as the oxygen tension and glucose levels. This is particularly important because little is currently known about small molecules, metabolites, and other factors that are required for the directed differentiation and/or long-term self-renewal of epithelial stem/progenitor cells in the lung. Indeed, identification of these molecules and factors, and their in vivo sources, is needed to improve organoid cultures. Interestingly, a remarkable progress has been made in the establishment of organoid cell culture that is based on a combination of several types of stromal cells. This progress will help in uncovering the interactions between these cell types in vivo and understanding how they are affected by inflammation, injury, and aging. Furthermore, other parameters such as the physical forces and extracellular matrix components, which most likely play critical roles in the control of stem cell behavior, are currently beginning to be explored using organoid cultures. However, more future studies are still need to clarify the functions of these parameters.

Finally, editing the genome using the CRISPR-Cas9 system is a powerful and advanced technology that still needs more studies in the field of lung stem cell behavior, particularly for the investigation of the functions of specific genes in stem cell behavior such as the regulatory genes of the self-renewal/proliferation or the differentiation of lung-specific stem cells. This technology is also important for modelling various pulmonary diseases in humans. Since CRISPR-Cas9-mediated *gene editing is more powerful in* hPSCs, the hPSC-derived organoids are powerful and advanced tools for different translationally based research applications such as the discovery of various drugs. However, the current major challenge and obstacle to overcome in the lung research is to develop methods that allow obtaining completely differentiated hPSC cultures into lung-specific cell types, such as *type I alveolar epithelial* cells. More research is still needed to overcome these obstacles and both identify and characterize more specific markers for the isolation and purification of lung-specific stem/progenitor cell populations in humans that will advance lung organoid culture technology. This will help address outstanding and novel questions in lung field, particularly in the field of human lung biology. Organoid cultures are, therefore, expected to become a highly significant and advanced tool for translationally based and basic research in the lung.

References

Abler, L. L., Mansour, S. L., & Sun, X. (2009). Conditional gene inactivation reveals roles for Fgf10 and Fgfr2 in establishing a normal pattern of epithelial branching in the mouse lung. *Developmental Dynamics, 238*, 1999–2013.

Adams, G. B., Chabner, K. T., Alley, I. R., et al. (2006). Stem cell engraftment at the endosteal niche is specified by the calcium-sensing receptor. *Nature, 439*, 599–603.

Adolfsson, J., Borge, O. J., Bryder, D., et al. (2001). Upregulation of Flt3 expression within the bone marrow Lin(−)Sca1(+)c-kit(+) stem cell compartment is accompanied by loss of self-renewal capacity. *Immunity, 15*, 659–669.

Ajduk, A., & Zernicka-Goetz, M. (2016). Polarity and cell division orientation in the cleavage embryo: from worm to human. *Molecular Human Reproduction, 22*, 691–703.

Akram, K. M., Patel, N., Spiteri, M. A., & Forsyth, N. R. (2016). Lung regeneration: Endogenous and exogenous stem cell mediated therapeutic approaches. *International Journal of Molecular Sciences, 17*, 128.

Alanis, D. M., Chang, D. R., Akiyama, H., et al. (2014). Two nested developmental waves demarcate a compartment boundary in the mouse lung. *Nature Communications, 5*, 3923.

Alder, J. K., Barkauskas, C. E., Limjunyawong, N., et al. (2015). Telomere dysfunction causes alveolar stem cell failure. *Proceedings of the National Academy of Sciences of the United States of America, 112*, 5099–5104.

Alexson, T. O., Hitoshi, S., Coles, B., Bernstein, A., & van der Kooy, D. (2006). Notch signaling is required to maintain all neural stem cell populations–irrespective of spatial or temporal niche. *Developmental Neuroscience, 28*, 34–48.

Ameis, D., Khoshgoo, N., Iwasiow, B. M., Snarr, P., & Keijzer, R. (2017). MicroRNAs in lung development and disease. *Paediatric Respiratory Reviews, 22*, 38–43.

Audet, J., Miller, C. L., Rose-John, S., Piret, J., & Eaves, C. (2001). Distinct role of gp130 activation in promoting self-renewal divisions by mitogenically stimulated murine hematopoietic stem cells. *Proceedings of the National Academy of Sciences of the United States of America, 98*, 1757–1762.

Balasooriya, G., Goschorska, M., Piddini, E., & Rawlins, E. L. (2017). FGFR2 is required for airway basal cell self-renewal and terminal differentiation. *Development, 144*, 1600–1606.

Balasubramaniam, V., Mervis, C., Maxey, A., Markham, N., & Abman, S. H. (2007). Hyperoxia reduces bone marrow, circulating, and lung endothelial progenitor cells in the developing lung: Implications for the pathogenesis of bronchopulmonary dysplasia. *American Journal of Physiology. Lung Cellular and Molecular Physiology, 292*, L1073–L1084.

Barkauskas, C. E., Cronce, M. J., Rackley, C. R., et al. (2013). Type 2 alveolar cells are stem cells in adult lung. *The Journal of Clinical Investigation, 123*, 3025–3036.

© Springer International Publishing AG, part of Springer Nature 2018
A. El-Hashash, *Lung Stem Cell Behavior*,
https://doi.org/10.1007/978-3-319-95279-6

Barkauskas, C. E., Chung, M.-I., Fioret, B., Gao, X., Katsura, H., & Hogan, B. L. (2017). Lung organoids: Current uses and future promise. *Development, 144*, 986–997.

Barker, N., Huch, M., Kujala, P., et al. (2010). Lgr5+ve stem cells drive self-renewal in the stomach and build long-lived gastric units in vitro. *Cell Stem Cell, 6*, 25–36.

Batard, P., Monier, M. N., Fortunel, N., et al. (2000). TGF-(beta)1 maintains hematopoietic immaturity by a reversible negative control of cell cycle and induces CD34 antigen up-modulation. *Journal of Cell Science, 113*, 383–390.

Bellusci, S., Henderson, R., Winnier, G., Oikawa, T., & Hogan, B. L. (1996). Evidence from normal expression and targeted misexpression that bone morphogenetic protein (Bmp-4) plays a role in mouse embryonic lung morphogenesis. *Development, 122*, 1693–1702.

Bellusci, S., Grindley, J., Emoto, H., Itoh, N., & Hogan, B. L. (1997). Fibroblast growth factor 10 (FGF10) and branching morphogenesis in the embryonic mouse lung. *Development, 124*(23), 4867–4878.

Berdnik, D., Török, T., González-Gaitán, M., & Knoblich, J. A. (2002). The endocytic protein alpha-Adaptin is required for numb-mediated asymmetric cell division in Drosophila. *Developmental Cell, 3*, 221–231.

Berika, M., Elgayyar, M., & El-Hashash, A. H. (2014). Asymmetric cell divisions of stem cells in the lung and other systems. *Frontiers in Cell and Development Biology, 2*, 33–42.

Berika, M., Ku, J., Huang, H., & El-Hashash, A. H. (2016). Gene and signals regulating stem cell fate. In A. El-Hashash (Ed.), *Developmental and stem cell biology in health and disease* (pp. 36–48). Madison: Bentham Science Publisher, USA.

Bertoncello, I. (2016). Properties of adult lung stem and progenitor cells. *Journal of Cellular Physiology, 231*, 2582–2589.

Betschinger, J., & Knoblich, J. A. (2004). Dare to be different: Asymmetric cell division in Drosophila, C. elegans and vertebrates. *Current Biology, 14*, R674–R685.

Bhaskaran, M., Wang, Y., Zhang, H., et al. (2009). MicroRNA-127 modulates fetal lung development. *Physiological Genomics, 37*, 268–278.

Blaisdell, C. J., Gail, D. B., & Nabel, E. G. (2009). National heart, lung, and blood institute perspective: Lung progenitor and stem cells--gaps in knowledge and future opportunities. *Stem Cells, 27*(9), 2263–2270.

Blank, U., & Karlsson, S. (2011). The role of Smad signaling in hematopoiesis and translational hematology. *Leukemia, 25*, 1379–1388.

Blanpain, C., Lowry, W. E., Pasolli, H. A., & Fuchs, E. (2006). Canonical notch signaling functions as a commitment switch in the epidermal lineage. *Genes & Development, 20*(21), 3022–3035.

Boitano, S., Safdar, Z., Welsh, D., Bhattacharya, J., & Koval, M. (2004). Cell-cell interactions in regulating lung function. *American Journal of Physiology. Lung Cellular and Molecular Physiology, 287*, L455–L459.

Borge, O. J., Ramsfjell, V., Veiby, O., Murphy, M. J., Jr., Lok, S., & Jacobsen, S. E. (1996). Thrombopoietin, but not erythropoietin promotes viability and inhibits apoptosis of multipotent murine hematopoietic progenitor cells in vitro. *Blood, 88*, 2859–2870.

Borthwick, D. W., Shahbazian, M., Krantz, Q. T., Dorin, J. R., & Randell, S. H. (2001). Evidence for stem-cell niches in the tracheal epithelium. *American Journal of Respiratory Cell and Molecular Biology, 24*, 662–670.

Bowie, M. B., McKnight, K. D., Kent, D. G., McCaffrey, L., Hoodless, P. A., & Eaves, C. J. (2006). Hematopoietic stem cells proliferate until after birth and show a reversible phase-specific engraftment defect. *The Journal of Clinical Investigation, 116*, 2808–2816.

Brawley, C., & Matunis, E. (2004). Regeneration of male germline stem cells by spermatogonial dedifferentiation in vivo. *Science, 304*, 1331–1334.

Broutier, L., Andersson-Rolf, A., Hindley, C. J., et al. (2016). Culture and establishment of self-renewing human and mouse adult liver and pancreas 3D organoids and their genetic manipulation. *Nature Protocols, 11*, 1724–1743.

Buckley, S., Driscoll, B., Anderson, K. D., & Warburton, D. (1997). Cell cycle in alveolar epithelial type II cells: Integration of Matrigel and KGF. *The American Journal of Physiology, 273*, L572–L580.

Buckley, S., Barsky, L., Driscoll, B., Weinberg, K., Anderson, K. D., & Warburton, D. (1998). Apoptosis and DNA damage in type 2 alveolar epithelial cells cultured from hyperoxic rats. *The American Journal of Physiology, 274*(5 Pt 1), L714–L720.

Buckley, S., Barsky, L., Weinberg, K., & Warburton, D. (2005). In vivo inosine protects alveolar epithelial type 2 cells against hyperoxia-induced DNA damage through MAP kinase signaling. *American Journal of Physiology. Lung Cellular and Molecular Physiology, 288*, L569–L575.

Buza-Vidas, N., Antonchuk, J., Qian, H., et al. (2006). Cytokines regulate postnatal hematopoietic stem cell expansion: Opposing roles of thrombopoietin and LNK. *Genes & Development, 20*, 2018–2023.

Cairns, J. M. (1975). The function of the ectodermal apical ridge and distinctive characteristics of adjacent distal mesoderm in the avian wing-bud. *Journal of Embryology and Experimental Morphology, 34*(1), 155–169.

Calvi, L. M., Adams, G. B., Weibrecht, K. W., et al. (2003). Osteoblastic cells regulate the haematopoietic stem cell niche. *Nature, 425*, 841–846.

Cao, Z., Lis, R., Ginsberg, M., et al. (2016). Targeting of the pulmonary capillary vascular niche promotes lung alveolar repair and ameliorates fibrosis. *Nature Medicine, 22*, 154–162.

Carraro, G., El-Hashash, A., Guidolin, D., et al. (2009). miR-17 family of microRNAs controls FGF10-mediated embryonic lung epithelial branching morphogenesis through MAPK14 and STAT3 regulation of E-Cadherin distribution. *Developmental Biology, 333*, 238–250.

Cayouette, M., & Raff, M. (2002). Asymmetric segregation of Numb: A mechanism for neural specification from Drosophila to mammals. *Nature Neuroscience, 5*, 1265–1269.

Cayouette, M., & Raff, M. (2003). The orientation of cell division influences cell-fate choice in the developing mammalian retina. *Development, 130*, 2329–2339.

Chapman, H. A., Li, X., Alexander, J. P., Brumwell, A., Lorizio, W., Tan, K., Sonnenberg, A., Wei, Y., & Vu, T. H. (2011). Integrin α6β4 identifies an adult distal lung epithelial population with regenerative potential in mice. *The Journal of Clinical Investigation, 121*(7), 2855–2862.

Chenn, A1., & McConnell, S. K. (1995). Cleavage orientation and the asymmetric inheritance of Notch1 immunoreactivity in mammalian neurogenesis. *Cell, 82*(4), 631–641.

Chen, F., & Fine, A. (2016). Stem cells in lung injury and repair. *The American Journal of Pathology, 186*, 2544–2550.

Chen, L., & Zosky, G. R. (2017). Lung development. *Photochemical & Photobiological Sciences, 16*, 339–346.

Chen, F., Desai, T. J., Qian, J., Niederreither, K., Lü, J., & Cardoso, W. V. (2007). Inhibition of Tgf beta signaling by endogenous retinoic acid is essential for primary lung bud induction. *Development, 134*, 2969–2979.

Chen, H., Matsumoto, K., Brockway, B. L., Rackley, C. R., Liang, J., Lee, J. H., Jiang, D., Noble, P. W., Randell, S. H., Kim, C. F., & Stripp, B. R. (2012a). Airway epithelial progenitors are region specifi c and show differential responses to bleomycin-induced lung injury. *Stem Cells, 30*(9), 1948–1960.

Chen, L., Acciani, T., Le Cras, T., Lutzko, C., & Perl, A. K. (2012b). Dynamic regulation of platelet-derived growth factor receptor alpha expression in alveolar fi broblasts during realveolarization. *American Journal of Respiratory Cell and Molecular Biology, 47*(4), 517–527.

Chen, K., Huang, Y., & Chen, J. (2013). Understanding and targeting cancer stem cells: Therapeutic implications and challenges. *Acta Pharmacologica Sinica, 34*, 732–740.

Chen, C., Fingerhut, J. M., & Yamashita, Y. M. (2016). The ins(ide) and outs(ide) of asymmetric stem cell division. *Current Opinion in Cell Biology, 43*, 1–6.

Christensen, J. L., & Weissman, I. L. (2001). Flk-2 is a marker in hematopoietic stem cell differentiation: A simple method to isolate long-term stem cells. *Proceedings of the National Academy of Sciences of the United States of America, 98*, 14541–14546.

Chuang, P. T., Kawcak, T., & McMahon, A. P. (2003). Feedback control of mammalian Hedgehog signaling by the Hedgehog-binding protein, Hip1, modulates Fgf signaling during branching morphogenesis of the lung. *Genes & Development, 17*, 342–347.

Clark, J., Alvarez, D. F., Alexeyev, M., King, J. A., Huang, L., Yoder, M. C., & Stevens, T. (2008). Regulatory role for nucleosome assembly protein-1 in the proliferative and vasculogenic phenotype of pulmonary endothelium. *American Journal of Physiology. Lung Cellular and Molecular Physiology, 294*(3), L431–L439.

Clevers, H. (2016). Modeling development and disease with organoids. *Cell, 165*, 1586–1597.

Clevers, H., Loh, K. M., & Nusse, R. (2014). An integral program for tissue renewal and regeneration: Wnt signaling and stem cell control. *Science, 346*, 1248012.

Cobas, M., Wilson, A., Ernst, B., et al. (2004). Beta-catenin is dispensable for hematopoiesis and lymphopoiesis. *The Journal of Experimental Medicine, 199*, 221–229.

Cole, B. B., Smith, R. W., Jenkins, K. M., Graham, B. B., Reynolds, P. R., & Reynolds, S. D. (2010). Tracheal basal cells: A facultative progenitor cell pool. *The American Journal of Pathology, 177*(1), 362–376.

Colvin, J. S., White, A. C., Pratt, S. J., & Ornitz, D. M. (2001). Lung hypoplasia and neonatal death in Fgf9-null mice identify this gene as an essential regulator of lung mesenchyme. *Development, 128*(11), 2095–2106.

Conboy, I. M., & Rando, T. A. (2002). The regulation of Notch signaling controls satellite cell activation and cell fate determination in postnatal myogenesis. *Developmental Cell, 3*(3), 397–409.

Conboy, M. J., Karasov, A. O., & Rando, T. A. (2007). High incidence of non-random template strand segregation and asymmetric fate determination in dividing stem cells and their progeny. *PLoS Biology, 5*(5), e102.

Croce, C. M. (2009). Causes and consequences of microRNA dysregulation in cancer. *Nature Reviews. Genetics, 10*(10), 704–714.

Curradi, G., Walters, M. S., Ding, B.-S., et al. (2012). Airway basal cell vascular endothelial growth factor mediated cross-talk regulates endothelial cell-dependent growth support of human airway basal cells. *Cellular and Molecular Life Sciences, 69*, 2217–2231.

Daynac, M., & Petritsch, C. K. (2017). Regulation of asymmetric cell division in mammalian neural stem and cancer precursor cells. In J. P. Tassan & J. Kubiak (Eds.), *Asymmetric cell division in development, differentiation and cancer. Results and problems in cell differentiation* (Vol. 61). Cham: Springer.

De Langhe, S. P., Carraro, G., Warburton, D., Hajihosseini, M. K., & Bellusci, S. (2006). Levels of mesenchymal FGFR2 signaling modulate smooth muscle progenitor cell commitment in the lung. *Developmental Biology, 299*, 52–62.

De Langhe, S. P., Carraro, G., Tefft, D., et al. (2008). Formation and differentiation of multiple mesenchymal lineages during lung development is regulated by beta-catenin signaling. *PLoS One, 3*, e1516.

del Moral, P.-M., & Warburton, D. (2010). Explant culture of mouse embryonic whole lung, isolated epithelium, or mesenchyme under chemically defined conditions as a system to evaluate the molecular mechanism of branching morphogenesis and cellular differentiation. *Methods in Molecular Biology, 633*, 71–79.

del Moral, P. M., De Langhe, S. P., Sala, F. G., et al. (2006). Differential role of FGF9 on epithelium and mesenchyme in mouse embryonic lung. *Developmental Biology, 293*, 77–89.

Delaney, C., Varnum-Finney, B., Aoyama, K., Brashem-Stein, C., & Bernstein, I. D. (2005). Dose-dependent effects of the Notch ligand Delta1 on ex vivo differentiation and in vivo marrow repopulating ability of cord blood cells. *Blood, 106*, 2693–2699.

Desai, T. J., Brownfield, D. G., & Krasnow, M. A. (2014). Alveolar progenitor and stem cells in lung development, renewal and cancer. *Nature*, 1–16.

Dewey, E. B., Taylor, D. T., & Johnston, C. A. (2015). Cell fate decision making through oriented cell division. *Journal of Developmental Biology, 3*, 129–157.

Ding, B.-S., Nolan, D. J., Guo, P., et al. (2011). Endothelial-derived inductive angiocrine signals initiate and sustain regenerative lung alveolarization. *Cell, 147*(3), 539–553.

Ding, Y., Zhao, R., Zhao, X., Matthay, M. A., Nie, H. G., & Ji, H. L. (2017). ENaCs as both effectors and regulators of MiRNAs in lung epithelial development and regeneration. *Cellular Physiology and Biochemistry, 44*, 1120–1132.

Doe, C. Q. (2008). Neural stem cells: Balancing self-renewal with differentiation. *Development, 135*, 1575–1587.

Driscoll, B., Buckley, S., Bui, K. C., Anderson, K. D., & Warburton, D. (2000). Telomerase in alveolar epithelial development and repair. *American Journal of Physiology. Lung Cellular and Molecular Physiology, 279*, L1191–L1198.

Drummond, M. L., & Prehoda, K. E. (2016). Molecular control of atypical Protein Kinase C: Tipping the balance between self-renewal and differentiation. *Journal of Molecular Biology, 428*, 1455–1464.

Du, Q1., Stukenberg, P. T., & Macara, I. G. (2001). A mammalian Partner of inscuteable binds NuMA and regulates mitotic spindle organization. *Nature Cell Biology, 3*(12), 1069–1075.

Duncan, A. W., Rattis, F. M., DiMascio, L. N., et al. (2005). Integration of Notch and Wnt signaling in hematopoietic stem cell maintenance. *Nature Immunology, 6*, 314–322.

Dye, B. R., Hill, D. R., Ferguson, M. A. H., et al. (2015). In vitro generation of human pluripotent stem cell derived lung organoids. *Elife, 4*, e05098.

Dye, B. R., Dedhia, P. H., Miller, A. J., et al. (2016). A bioengineered niche promotes in vivo engraftment and maturation of pluripotent stem cell derived human lung organoids. *Elife, 5*, e19732.

Eblaghie, M. C., Reedy, M., Oliver, T., Mishina, Y., & Hogan, B. L. (2006). Evidence that autocrine signaling through Bmpr1a regulates the proliferation, survival and morphogenetic behavior of distal lung epithelial cells. *Developmental Biology, 291*, 67–82.

El Agha, E., Herold, S., Al Alam, D., et al. (2014). Fgf10-positive cells represent a progenitor cell population during lung development and postnatally. *Development, 141*, 296–306.

El-Badrawy, M. K., Shalabi, N. M., Mohamed, M. A., Ragab, A., & Abdelwahab, H. W. (2016). Stem cells and lung regeneration. *International Journal of Stem Cells, 9*, 31–35.

El-Hashash, A. H. (2013). Lung stem cells: Mechanisms of behavior, development and regeneration. *Anatomy and Physiology, 3*, 119–128.

El-Hashash, A. H. (2015). Asymmetric cell divisions of stem/progenitor cells. In D. David Warburton (Ed.), *Stem cells, tissue engineering and regenerative medicine* (pp. 41–58). New Jersey: World Scientific Publishing.

El-Hashash, A. H. (2016a). Stem cells, developmental biology and reparative/regenerative medicine: Tools and hope for better human life. In A. El-Hashash (Ed.), *Developmental and stem cell biology in health and disease* (pp. 3–5). Madison: Bentham Science Publisher, USA.

El-Hashash, A. H. (2016b). Neural crest stem cells: A hope for neural regeneration. In E. Abdelalim (Ed.), *Recent advances in stem cells: From basic research to clinical applications* (pp. 233–250). Berlin: Springer Science Publisher.

El-Hashash, A. H. (2018). Intrinsic vs extrinsic intrinsic regulatory mechanisms of lung stem cell biology and behavior. *Journal of Stem Cells, 12*, 187–190.

El-Hashash, A. H., & Warburton, D. (2011). Cell polarity and spindle orientation in the distal epithelium of embryonic lung. *Developmental Dynamics, 240*, 441–445.

El-Hashash, A. H., & Warburton, D. (2012). Numb expression and asymmetric versus symmetric cell division in embryonic distal lung epithelium. *Journal of Histochemistry and Cytochemistry, 60*, 675–682.

El-Hashash, A. H., Al Alam, D., Turcatel, G., Bellusci, S., & Warburton, D. (2011a). Eyes absent 1 (Eya1) is a critical coordinator of epithelial, mesenchymal and vascular morphogenesis in the mammalian lung. *Developmental Biology, 350*, 112–126.

El-Hashash, A. H., Al Alam, D., Turcatel, G., et al. (2011b). Six1 transcription factor is critical for coordination of epithelial, mesenchymal and vascular morphogenesis in the mammalian lung. *Developmental Biology, 353*, 242–258.

El-Hashash, A., Turcatel, G., Alam, D., Buckley, S., Bellusci, S., & Warburton, D. (2011c). Eya1 controls cell polarity, spindle orientation, cell fate and Notch signaling in distal embryonic lung epithelium. *Development, 138*, 1395–1407.

Elshahawy, S., Ibrahim, A., Soliman, S., Berika, M., & El-Hashash, A. H. (2016). Behavior and asymmetric cell divisions of stem cells. In A. El-Hashash (Ed.), *Developmental and stem cell biology in health and disease* (pp. 81–104). Madison: Bentham Science Publisher, USA.

Ema, H., Sudo, K., Seita, J., et al. (2005). Quantification of self-renewal capacity in single hematopoietic stem cells from normal and Lnk-deficient mice. *Developmental Cell, 8*, 907–914.

Engelhardt, J. F. (2001). Stem cell niches in the mouse airway. *American Journal of Respiratory Cell and Molecular Biology, 24*, 649–652.

Enver, T., Heyworth, C. M., & Dexter, T. M. (1998). Do stem cells play dice? *Blood, 92*, 348–351.

Evans, M. J., Cabral, L. J., Stephens, R. J., & Freeman, G. (1975). Transformation of alveolar type 2 cells to type 1 cells following exposure to NO2. *Experimental and Molecular Pathology, 22*, 142–150.

Evans, M. J., Shami, S. G., Cabral-Anderson, L. J., & Dekker, N. P. (1986). Role of nonciliated cells in renewal of the bronchial epithelium of rats exposed to NO2. *The American Journal of Pathology, 123*(1), 126–133.

Evans, M. J., Van Winkle, L. S., Fanucchi, M. V., & Plopper, C. G. (2001). Cellular and molecular characteristics of basal cells in airway epithelium. *Experimental Lung Research, 27*(5), 401–415.

Fan, T., Wang, W., Zhang, B., et al. (2016). Regulatory mechanisms of microRNAs in lung cancer stem cells. *SpringerPlus, 5*(1), 1762.

Firth, A. L., Dargitz, C. T., Qualls, S. J., et al. (2014). Generation of multiciliated cells in functional airway epithelia from human induced pluripotent stem cells. *Proceedings of the National Academy of Sciences of the United States of America, 111*(17), E1723–E1730.

Florian, M. C., & Geiger, H. (2010). Concise review: Polarity in stem cells, disease, and aging. *Stem Cells (Dayton, Ohio), 28*(9), 1623–1629.

Forostyak, O., Romanyuk, N., Verkhratsky, A., Sykova, E., & Dayanithi, G. (2013). Plasticity of calcium signaling cascades in human embryonic stem cell-derived neural precursors. *Stem Cells and Development, 22*(10), 1506–1521.

Fox, E., Shojaie, S., Wang, J., et al. (2015). Three-dimensional culture and FGF signaling drive differentiation of murine pluripotent cells to distal lung epithelial cells. *Stem Cells and Development, 24*, 21–35.

Frank, D. B., Peng, T., Zepp, J., et al. (2016). Emergence of a wave of Wnt signaling that regulates lung alveologenesis through controlling epithelial self-renewal and differentiation. *Cell Reports, 17*, 2312–2325.

Frise, E., Knoblich, J. A., Younger-Shepherd, S., et al. (1996). The Drosophila Numb protein inhibits signaling of the Notch receptor during cell-cell interaction in sensory organ lineage. *Proceedings of the National Academy of Sciences of the United States of America, 93*, 11925–11932.

Fu, Y., Li, H., & Hao, X. (2017). The self-renewal signaling pathways utilized by gastric cancer stem cells. *Tumor Biology*, 1–7.

Fuchs, E1., & Raghavan, S. (2002). Getting under the skin of epidermal morphogenesis. *Nature Reviews. Genetics, 3*(3), 199–209.

Fulcher, M. L., & Randell, S. H. (2013). Human nasal and tracheobronchial respiratory epithelial cell culture. *Methods in Molecular Biology, 945*, 109–121.

Gao, X., Vockley, C. M., Pauli, F., et al. (2013). Evidence for multiple roles for grainyheadlike 2 in the establishment and maintenance of human mucociliary airway epithelium. *Proceedings of the National Academy of Sciences of the United States of America, 110*, 9356–9351.

Gao, X., Bali, A. S., Randell, S. H., & Hogan, B. L. M. (2015). GRHL2 coordinates regeneration of a polarized mucociliary epithelium from basal stem cells. *The Journal of Cell Biology, 211*, 669–682.

Garbe, A., Spyridonidis, A., Mobest, D., et al. (1997). Transforming growth factor-beta 1 delays formation of granulocyte-macrophage colony-forming cells, but spares more primitive progenitors during ex vivo expansion of CD34+ haemopoietic progenitor cells. *British Journal of Haematology, 99*, 951–958.

Ghaedi, M., Calle, E. A., Mendez, J. J., et al. (2013). Human iPS cell derived alveolar epithelium repopulates lung extracellular matrix. *The Journal of Clinical Investigation, 123*, 4950–4962.

Giangreco, A., Reynolds, S. D., & Stripp, B. R. (2002). Terminal bronchioles harbor a unique airway stem cell population that localizes to the bronchoalveolar duct junction. *The American Journal of Pathology, 161*, 173–182.

Giangreco, A., Shen, H., Reynolds, S. D., et al. (2004). Molecular phenotype of airway side population cells. *American Journal of Physiology. Lung Cellular and Molecular Physiology, 286*, L624–L630.

Giebel, B., & Wodarz, A. (2012). Notch signaling: Numb makes the difference. *Curr Biol, 22*, R133–R135.

Goldstein, B., & Macara, I. G. (2007). The PAR proteins: Fundamental players in animal cell polarization. *Developmental Cell, 13*, 609–622.

Gómez-López, S., Lerner, R. G., & Petritsch, C. (2014). Asymmetric cell division of stem and progenitor cells during homeostasis and cancer. *Cellular and Molecular Life Sciences, 71*, 575–597.

Gönczy, P. (2008). Mechanisms of asymmetric cell division: Flies and worms pave the way. *Nature Reviews. Molecular Cell Biology, 9*(5), 355–366.

Goss, A. M., Tian, Y., Tsukiyama, T., et al. (2009). Wnt2/2b and beta-catenin signaling are necessary and sufficient to specify lung progenitors in the foregut. *Developmental Cell, 17*, 290–298.

Goss, A. M., Tian, Y., Cheng, L., et al. (2011). Wnt2 signaling is necessary and sufficient to activate the airway smooth muscle program in the lung by regulating myocardin/Mrtf-B and Fgf10 expression. *Developmental Biology, 356*, 541–552.

Gotoh, S., Ito, I., Nagasaki, T., et al. (2014). Generation of alveolar epithelial spheroids via isolated progenitor cells from human pluripotent stem cells. *Stem Cell Reports, 3*, 394–403.

Green, M. D., Huang, S. X., & Snoeck, H. W. (2013). Stem cells of the respiratory system: From identification to differentiation into functional epithelium. *BioEssays, 35*, 261–270.

Greggio, C., De Franceschi, F., Figueiredo-Larsen, M., et al. (2013). Artificial three-dimensional niches deconstruct pancreas development in vitro. *Development, 140*, 4452–4462.

Greggio, C., De Franceschi, F., Figueiredo-Larsen, M., & Grapin-Botton, A. (2014). In vitro pancreas organogenesis from dispersed mouse embryonic progenitors. *Journal of Visualized Experiments, 89*, e51725.

Greggio, C., De Franceschi, F., & Grapin-Botton, A. (2015). Concise reviews: In vitro-produced pancreas organogenesis models in three dimensions: Self-organization from few stem cells or progenitors. *Stem Cells, 33*, 8–14.

Gulino, A., Di Marcotullio, L., & Screpanti, I. (2010). The multiple functions of Numb. *Experimental Cell Research, 316*, 900–906.

Guo, M., Jan, L. Y., & Jan, Y. N. (1996). Control of daughter cell fates during asymmetric division: Interaction of Numb and Notch. *Neuron, 17*, 27–41.

Gupte, V. V., Ramasamy, S. K., Reddy, R., et al. (2009). Overexpression of fibroblast growth factor-10 during both inflammatory and fibrotic phases attenuates bleomycin-induced pulmonary fibrosis in mice. *American Journal of Respiratory and Critical Care Medicine, 180*, 424–436.

Gurdon, J. B. (1988). A community effect in animal development. *Nature, 336*, 772–774.

Guseh, J. S., Bores, S. A., Stanger, B. Z., et al. (2009). Notch signaling promotes airway mucous metaplasia and inhibits alveolar development. *Development, 136*, 1751–1759.

Hackett, T. L., Shaheen, F., Johnson, A., et al. (2008). Characterization of side population cells from human airway epithelium. *Stem Cells, 26*, 2576–2585.

Harris-Johnson, K. S., Domyan, E. T., Vezina, C. M., et al. (2009). beta-Catenin promotes respiratory progenitor identity in mouse foregut. *Proceedings of the National Academy of Sciences of the United States of America, 106*, 16287–16292.

Haydar, T. F., Ang, E., Jr., & Rakic, P. (2003). Mitotic spindle rotation and mode of cell division in the developing telencephalon. *Proceedings of the National Academy of Sciences of the United States of America, 100*, 2890–2895.

Hegab, A. E., Ha, V. L., Gilbert, J. L., et al. (2011). Novel stem/progenitor cell population from murine tracheal submucosal gland ducts with multipotent regenerative potential. *Stem Cells, 29*, 1283–1293.

Hegab, A. E., Arai, D., Gao, J., et al. (2015). Mimicking the niche of lung epithelial stem cells and characterization of several effectors of their in vitro behavior. *Stem Cell Research, 15*, 109–121.

Herriges, M., & Morrisey, E. E. (2014). Lung development: Orchestrating the generation and regeneration of a complex organ. *Development, 141*, 502–513.

Hill, A. R., Donaldson, J. E., Blume, C., et al. (2016). IL-1α mediates cellular cross-talk in the airway epithelial mesenchymal trophic unit. *Tissue Barriers, 4*, e1206378.

Hogan, B. L. (1999). Morphogenesis. *Cell, 96*(2), 225–233.

Hogan, B. L. M., Barkauskas, C. E., Chapman, H. A., et al. (2014). Repair and regeneration of the respiratory system: Complexity, plasticity, and mechanisms of lung stem cell function. *Cell Stem Cell, 15*, 123–138.

Homem, C. C., & Knoblich, J. A. (2012). Drosophila neuroblasts: A model for stem cell biology. *Development, 139*, 4297–4310.

Hong, K. U., Reynolds, S. D., Watkins, S., et al. (2004). Basal cells are a multipotent progenitor capable of renewing the bronchial epithelium. *The American Journal of Pathology, 164*, 577–588.

Hsu, Y. C., Osinski, J., Campbell, C. E., et al. (2011). Mesenchymal nuclear factor I B regulates cell proliferation and epithelial differentiation during lung maturation. *Developmental Biology, 354*, 242–252.

Huang, S. X. L., Islam, M. N., O'Neill, J., et al. (2014). Efficient generation of lung and airway epithelial cells from human pluripotent stem cells. *Nature Biotechnology, 32*(1), 84–91.

Huang, S. X. L., Green, M. D., de Carvalho, A. T., et al. (2015). The in vitro generation of lung and airway progenitor cells from human pluripotent stem cells. *Nature Protocols, 10*, 413–425.

Huch, M., Bonfanti, P., Boj, S. F., et al. (2013a). Unlimited in vitro expansion of adult bi-potent pancreas progenitors through the Lgr5/Rspondin axis. *The EMBO Journal, 32*, 2708–2721.

Huch, M., Dorrell, C., Boj, S. F., et al. (2013b). In vitro expansion of single Lgr5+ liver stem cells induced by Wnt-driven regeneration. *Nature, 494*, 247–250.

Huch, M., Gehart, H., van Boxtel, R., et al. (2015). Long-term culture of genome-stable bipotent stem cells from adult human liver. *Cell, 160*, 299–312.

Huttner, W. B., & Kosodo, Y. (2005). Symmetric versus asymmetric cell division during neurogenesis in the developing vertebrate central nervous system. *Current Opinion in Cell Biology, 17*, 648–657.

Hynds, R. E., Butler, C. R., Janes, S. M., & Giangreco, A. (2016). Expansion of human airway basal stem cells and their differentiation as 3D tracheospheres. *Methods in Molecular Biology, 1*, 11.

Ibrahim, A., & El-Hashash, A. H. (2015). Lung stem cell behavior in development and regeneration. *Edorium Journal of Stem Cell Research and Therapy, 1*, 1–13.

Ikuta, K., & Weissman, I. L. (1992). Evidence that hematopoietic stem cells express mouse c-kit but do not depend on steel factor for their generation. *Proceedings of the National Academy of Sciences of the United States of America, 89*, 1502–1506.

Irwin, D., Helm, K., Campbell, N., et al. (2007). Neonatal lung side population cells demonstrate endothelial potential and are altered in response to hyperoxia-induced lung simplification. *American Journal of Physiology. Lung Cellular and Molecular Physiology, 293*, L941–L951.

Ito, K., Hirao, A., Arai, F., et al. (2004). Regulation of oxidative stress by ATM is required for self-renewal of haematopoietic stem cells. *Nature, 431*, 997–1002.

Ito, M., Liu, Y., Yang, Z., et al. (2005). Stem cells in the hair follicle bulge contribute to wound repair but not to homeostasis of the epidermis. *Nature Medicine, 11*, 1351–1354.

Ito, K., Hirao, A., Arai, F., et al. (2006). Reactive oxygen species act through p38 MAPK to limit the lifespan of hematopoietic stem cells. *Nature Medicine, 12*, 446–451.

Itoh, F., Itoh, S., Goumans, M. J., et al. (2004). Synergy and antagonism between Notch and BMP receptor signaling pathways in endothelial cells. *The EMBO Journal, 23*, 541–551.

Jackson, S.-R., Lee, J., Reddy, R., Williams, G. N., Kikuchi, A., Freiberg, Y., Warburton, D., & Driscoll, B. (2011). Partial pneumonectomy of telomerase null mice carrying shortened telomeres initiates cell growth arrest resulting in a limited compensatory growth response. *American Journal of Physiology. Lung Cellular and Molecular Physiology, 300*(6), L898–L909.

Jain, R, Barkauskas, CE, Takeda, N, Bowie, EJ, Aghajanian, H, Wang, Q, Padmanabhan, A, Manderfield, LJ, Gupta, M, Li, D, Li, L, Trivedi, CM, Hogan, BL, Epstein, JA. (2015). Plasticity of Hopx(+) type I alveolar cells to regenerate type II cells in the lung. *Nat Commun, 6*, 6727–6737.

Jakkula, M., Le Cras, T. D., Gebb, S., Hirth, K. P., Tuder, R. M., Voelkel, N. F., & Abman, S. H. (2000). Inhibition of angiogenesis decreases alveolarization in the developing rat lung. *American Journal of Physiology. Lung Cellular and Molecular Physiology, 279*(3), L600–L607.

Jaskoll, T. F., Don-Wheeler, G., Johnson, R., & Slavkin, H. C. (1988). Embryonic mouse lung morphogenesis and type II cytodifferentiation in serumless, chemically defined medium using prolonged in vitro cultures. *Cell Differentiation, 24*, 105–117.

Jiang, J. X., & Li, L. (2009). Potential therapeutic application of adult stem cells in acute respiratory distress syndrome. *Chinese Journal of Traumatology, 12*, 228–233.

Juven-Gershon, T., Shifman, O., Unger, T., et al. (1998). The Mdm2 oncoprotein interacts with the cell fate regulator Numb. *Molecular and Cellular Biology, 18*, 3974–3982.

Kai, T., & Spradling, A. (2004). Differentiating germ cells can revert into functional stem cells in Drosophila melanogaster ovaries. *Nature, 428*, 564–569.

Karanu, F. N., Murdoch, B., Gallacher, L., et al. (2000). The notch ligand jagged-1 represents a novel growth factor of human hematopoietic stem cells. *The Journal of Experimental Medicine, 192*, 1365–1372.

Karanu, F. N., Murdoch, B., Miyabayashi, T., et al. (2001). Human homologues of Delta-1 and Delta-4 function as mitogenic regulators of primitive human hematopoietic cells. *Blood, 97*, 1960–1967.

Karthaus, W. R., Iaquinta, P. J., Drost, J., et al. (2014). Identification of multipotent luminal progenitor cells in human prostate organoid cultures. *Cell, 159*, 163–175.

Kaushansky, K., & Drachman, J. G. (2002). The molecular and cellular biology of thrombopoietin: The primary regulator of platelet production. *Oncogene, 21*, 3359–3367.

Kim, N., & Vu, T. H. (2006). Parabronchial smooth muscle cells and alveolar myofibroblasts in lung development. *Birth Defects Research. Part C, Embryo Today, 78*, 80–89.

Kim, C. F., Jackson, E. L., Woolfenden, A. E., et al. (2005). Identification of bronchioalveolar stem cells in normal lung and lung cancer. *Cell, 121*, 823–835.

Kimura, S., Hara, Y., Pineau, T., et al. (1996). The T/ebp null mouse: Thyroid-specific enhancer-binding protein is essential for the organogenesis of the thyroid, lung, ventral forebrain, and pituitary. *Genes & Development, 10*, 60–69.

Kimura, S., Roberts, A. W., Metcalf, D., et al. (1998). Hematopoietic stem cell deficiencies in mice lacking c-Mpl, the receptor for thrombopoietin. *Proceedings of the National Academy of Sciences of the United States of America, 95*, 1195–1200.

Kimura, S., Ward, J. M., & Minoo, P. (1999). Thyroid-specific enhancer-binding protein/thyroid transcription factor 1 is not required for the initial specification of the thyroid and lung primordia. *Biochimie, 81*, 321–327.

Kirstetter, P., Anderson, K., Porse, B. T., et al. (2006). Activation of the canonical Wnt pathway leads to loss of hematopoietic stem cell repopulation and multilineage differentiation block. *Nature Immunology, 7*, 1048–1056.

Knoblich, J. A. (2001). Asymmetric cell division during animal development. *Nature Reviews. Molecular Cell Biology, 2*, 11–20.

Knoblich, J. A. (2010). Asymmetric cell division: Recent developments and their implications for tumour biology. *Nature Reviews. Molecular Cell Biology, 11*, 849–860.

Kohwi, M., & Doe, C. Q. (2013). Temporal fate specification and neural progenitor competence during development. *Nature Reviews. Neuroscience, 14*, 823–838.

Konno, D., Shioi, G., Shitamukai, A., Mori, A., Kiyonari, H., Miyata, T., & Matsuzaki, F. (2008). Neuroepithelial progenitors undergo LGN-dependent planar divisions to maintain self-renewability during mammalian neurogenesis. *Nature Cell Biology, 10*(1), 93–101.

Konishi, S., Gotoh, S., Tateishi, K., et al. (2016). Directed induction of functional multi-ciliated cells in proximal airway epithelial spheroids from human pluripotent stem cells. *Stem Cell Reports, 6*, 18–25.

Kosodo, Y., Röper, K., Haubensak, W., et al. (2004). Asymmetric distribution of the apical plasma membrane during neurogenic divisions of mammalian neuroepithelial cells. *The EMBO Journal, 23*, 2314–2324.

Kotton, D. N., & Morrisey, E. E. (2014). Lung regeneration: Mechanisms, applications and emerging stem cell populations. *Nature Medicine, 20*, 822–832.

Ku, J., & El-Hashash, A. H. (2016). Molecular control of the mode of cell division: A view from mammalian lung epithelial stem cells. *A Journal of Anatomy, 3*, 3–6.

Kuang, S., Kuroda, K., Le Grand, F., & Rudnicki, M. A. (2007). Asymmetric self-renewal and commitment of satellite stem cells in muscle. *Cell, 129*(5), 999–1010.

Kugler, M. C., Joyner, A. L., Loomis, C. A., & Munger, J. S. (2015). Sonic hedgehog signaling in the lung. From development to disease. *American Journal of Respiratory Cell and Molecular Biology., 52*(1), 1–13.

Kumar, P. A., Hu, Y., Yamamoto, Y., et al. (2011). Distal airway stem cells yield alveoli in vitro and during lung regeneration following H1N1 influenza infection. *Cell, 147*, 525–538.

Kunisato, A., Chiba, S., Nakagami-Yamaguchi, E., et al. (2003). HES-1 preserves purified hematopoietic stem cells ex vivo and accumulates side population cells in vivo. *Blood, 101*, 1777–1783.

Labbé, E., Letamendia, A., & Attisano, L. (2000). Association of Smads with lymphoid enhancer binding factor 1/T cell-specific factor mediates cooperative signaling by the transforming growth factor-beta and wnt pathways. *Proceedings of the National Academy of Sciences of the United States of America, 97*, 8358–8363.

Lama, V. N., & Phan, S. H. (2006). The extrapulmonary origin of fibroblasts: Stem/progenitor cells and beyond. *Proceedings of the American Thoracic Society, 3*(4), 373–376.

Lama, V. N., Harada, H., Badri, L. N., Flint, A., Hogaboam, C. M., McKenzie, A., Martinez, F. J., Toews, G. B., Moore, B. B., & Pinsky, D. J. (2006). Obligatory role for interleukin-13 in obstructive lesion development in airway allografts. *The American Journal of Pathology, 169*, 47–60.

Laresgoiti, U., Nikolić, M. Z., Rao, C., Brady, J. L., et al. (2016). Lung epithelial tip progenitors integrate glucocorticoid- and STAT3-mediated signals to control progeny fate. *Development, 143*, 3686–3699.

Larsson, J., Blank, U., Helgadottir, H., et al. (2003). TGF-beta signaling-deficient hematopoietic stem cells have normal self-renewal and regenerative ability in vivo despite increased proliferative capacity in vitro. *Blood, 102*, 3129–3135.

Larsson, J., Blank, U., Klintman, J., Magnusson, M., & Karlsson, S. (2005). Quiescence of hematopoietic stem cells and maintenance of the stem cell pool is not dependent on TGF-beta signaling in vivo. *Experimental Hematology, 33*, 592–596.

Lechler, T., & Fuchs, E. (2005). Asymmetric cell divisions promote stratification and differentiation of mammalian skin. *Nature, 437*, 275–280.

Lee, J.-H., Bhang, D. H., Beede, A., et al. (2014). Lung stem cell differentiation in mice directed by endothelial cells via a BMP4-NFATc1-thrombospondin-1 axis. *Cell, 156*, 440–455.

Lehmann, M., Baarsma, H. A., & Königshoff, M. (2016). WNT signaling in lung aging and disease. *Annals of the American Thoracic Society, 13*, S411–S416.

Li, C., Xiao, J., Hormi, K., et al. (2002). Wnt5a participates in distal lung morphogenesis. *Developmental Biology, 248*, 68–81.

Li, F., He, J., Wei, J., Cho, W. C., & Liu, X. (2015). Diversity of epithelial stem cell types in adult lung. *Stem Cells International, 2015*, 728307.

Liu, X., & Engelhardt, J. F. (2008). The glandular stem/progenitor cell niche in airway development and repair. *Proceedings of the American Thoracic Society, 5*(6), 682–688.

Liu, X., Driskell, R. R., & Engelhardt, J. F. (2007). Stem cells in the lung. *Methods in Enzymology*. Author manuscript; Available in PMC 2007 Feb 22. Published in final edited form as: Methods Enzymol. 2006, *419*, 285–321.

Liu, Y., & Hogan, B. L. (2002). Differential gene expression in the distal tip endoderm of the embryonic mouse lung. *Gene Expression Patterns, 2*, 229–233.

Liu, C., Glasser, S. W., Wan, H., et al. (2002). GATA-6 and thyroid transcription factor-1 directly interact and regulate surfactant protein-C gene expression. *The Journal of Biological Chemistry, 277*, 4519–4525.

Liu, K., Lin, Q., Wei, Y., et al. (2015). Gαs regulates asymmetric cell division of cortical progenitors by controlling Numb mediated Notch signaling suppression. *Neuroscience Letters, 597*, 97–103.

Liu, Y., Jiang, B.-J., Zhao, R.-Z., et al. (2016). Epithelial sodium channels in pulmonary epithelial progenitor and stem cells. *International Journal of Biological Sciences, 12*, 1150–1154.

Longmire, T. A., Ikonomou, L., Hawkins, F., et al. (2012). Efficient derivation of purified lung and thyroid progenitors from embryonic stem cells. *Cell Stem Cell, 10*, 398–411.

Lu, J., & Clark, A. G. (2012). Impact of microRNA regulation on variation in human gene expression. *Genome Research, 22*(7), 1243–1254.

Lu, Y., Thomson, J. M., Wong, H. Y., Hammond, S. M., & Hogan, B. L. (2007). Transgenic overexpression of the microRNA miR-17-92 cluster promotes proliferation and inhibits differentiation of lung epithelial progenitor cells. *Developmental Biology, 310*, 442–453.

Lu, Y., Okubo, T., Rawlins, E., et al. (2008). Epithelial progenitor cells of the embryonic lung and the role of microRNAs in their proliferation. *Proceedings of the American Thoracic Society, 5*, 300–304.

Lüdtke, T. H., Farin, H. F., Rudat, C., et al. (2013). Tbx2 controls lung growth by direct repression of the cell cycle inhibitor genes Cdkn1a and Cdkn1b. *PLoS Genetics, 9*, e1003189.

Macara, I. G. (2004a). Par proteins: Partners in polarization. *Current Biology, 14*(4), R160–R162.

Macara, I. G. (2004b). Parsing the polarity code. *Nature Reviews. Molecular Cell Biology, 5*(3), 220–231.

Mailleux, A. A., Kelly, R., Veltmaat, J. M., et al. (2005). Fgf10 expression identifies parabronchial smooth muscle cell progenitors and is required for their entry into the smooth muscle cell lineage. *Development, 132*, 2157–2166.

Mancini, S. J., Mantei, N., Dumortier, A., et al. (2005). Jagged1-dependent Notch signaling is dispensable for hematopoietic stem cell self-renewal and differentiation. *Blood, 105*, 2340–2342.

Marieb, E. N., & Keller, S. M. (2017). *Essentials of Human anatomy & physiology*, Books a la Carte Edition (12th ed.). Pearson. 656 page. Cambridge, England, UK.

Matsuzaki, Y., Kinjo, K., Mulligan, R. C., et al. (2004). Unexpectedly efficient homing capacity of purified murine hematopoietic stem cells. *Immunity, 20*, 87–93.

McCracken, K. W., Catá, E. M., Crawford, C. M., et al. (2014). Modelling human development and disease in pluripotent stem-cell-derived gastric organoids. *Nature, 516*, 400–404.

McDowell, E. M., Newkirk, C., & Coleman, B. (1985). Development of hamster tracheal epithelium: II. Cell proliferation in the fetus. *The Anatomical Record, 213*, 448–456.

McQualter, J. L., Yuen, K., Williams, B., & Bertoncello, I. (2010). Evidence of an epithelial stem/progenitor cell hierarchy in the adult mouse lung. *Proceedings of the National Academy of Sciences of the United States of America, 107*(4), 1414–1419.

Metcalf, D. (1993). Hematopoietic regulators: Redundancy or subtlety? *Blood, 82*, 3515–3523.

Micchelli, C. A., & Perrimon, N. (2006). Evidence that stem cells reside in the adult Drosophila midgut epithelium. *Nature, 439*(7075), 475–479.

Miller, B. H., & Wahlestedt, C. (2010). MicroRNA dysregulation in psychiatric disease. *Brain Research, 1338*, 89–99.

Miller, C. L., & Eaves, C. J. (1997). Expansion in vitro of adult murine hematopoietic stem cells with transplantable lympho-myeloid reconstituting ability. *Proceedings of the National Academy of Sciences of the United States of America, 94*, 13648–13653.

Mills, A. A., Zheng, B., Wang, X. J., Vogel, H., Roop, D. R., & Bradley, A. (1999). p63 is a p53 homologue required for limb and epidermal morphogenesis. *Nature, 398*(6729), 708–713.

Milner, L. A., Kopan, R., Martin, D. I., et al. (1994). A human homologue of the Drosophila developmental gene, Notch, is expressed in CD34+ hematopoietic precursors. *Blood, 83*, 2057–2062.

Molofsky, A. V., Pardal, R., & Morrison, S. J. (2004). Diverse mechanisms regulate stem cell self-renewal. *Current Opinion in Cell Biology, 16*(6), 700–707.

Mondrinos, M. J., Koutzaki, S., Lelkes, P. I., & Finck, C. M. (2007). A tissue engineered model of fetal distal lung tissue. *American Journal of Physiology. Lung Cellular and Molecular Physiology, 293*, L639–L650.

Mondrinos, M. J., Jones, P. L., Finck, C. M., & Lelkes, P. I. (2014). Engineering de novo assembly of fetal pulmonary organoids. *Tissue Engineering. Part A, 20*, 2892–2907.

Mori, M., Mahoney, J. E., Stupnikov, M. R., et al. (2015). Notch3-Jagged signaling controls the pool of undifferentiated airway progenitors. *Development, 142*, 258–267.

Morimoto, M., Liu, Z., Cheng, H. T., et al. (2010). Canonical Notch signaling in the developing lung is required for determination of arterial smooth muscle cells and selection of Clara versus ciliated cell fate. *Journal of Cell Science, 123*, 213–224.

Morrisey, E. E., & Hogan, B. L. M. (2010). Preparing for the first breath: Genetic and cellular mechanisms in lung development. *Developmental Cell, 18*, 8–23.

Morin, X., Jaouen, F., & Durbec, P. (2007). Control of planar divisions by the G-protein regulator LGN maintains progenitors in the chick neuroepithelium. *Nature Neuroscience, 10*(11), 1440–1448.

Morrison, S. J., & Kimble, J. (2006). Asymmetric and symmetric stem-cell divisions in development and cancer. Nature. *441*(7097), 1068–1074.

Morrison, S. J., & Scadden, D. T. (2014). The bone marrow niche for haematopoietic stem cells. *Nature, 505*, 327–334.

Morrison, S. J., & Weissman, I. L. (1994). The long-term repopulating subset of hematopoietic stem cells is deterministic and isolatable by phenotype. *Immunity, 1*, 661–673.

Mou, H., Zhao, R., Sherwood, R., et al. (2012). Generation of multipotent lung and airway progenitors from mouse ESCs and patient-specific cystic fibrosis iPSCs. *Cell Stem Cell, 10*, 385–397.

Mou, H., Vinarsky, V., Tata, P. R., et al. (2016). Dual SMAD signaling inhibition enables long-term expansion of diverse epithelial basal cells. *Cell Stem Cell, 19*, 217–231.

Nadkarni, R. R., Abed, S., & Draper, J. S. (2016). Organoids as a model system for studying human lung development and disease. *Biochemical and Biophysical Research Communications, 473*, 675–682.

Nandurkar, H. H., Robb, L., Tarlinton, D., et al. (1997). Adult mice with targeted mutation of the interleukin-11 receptor (IL11Ra) display normal hematopoiesis. *Blood, 90*, 2148–2159.

Nelson, W. J. (2003a). Epithelial cell polarity from the outside looking in. *News in Physiological Sciences, 18*, 143–146.

Nelson, W. J. (2003b). Adaptation of core mechanisms to generate cell polarity. *Nature, 422*, 766–774.

Neumüller, R. A., & Knoblich, J. A. (2009). Dividing cellular asymmetry: Asymmetric cell division and its implications for stem cells and cancer. *Genes & Development, 23*(23), 2675–2699.

Nikolić, M. Z., & Rawlins, E. L. (2017). Lung organoids and their use to study cell-cell interaction. *Current Pathobiology Reports, 5*, 223.

Nilsson, S. K., Johnston, H. M., Whitty, G. A., et al. (2005). Osteopontin, a key component of the hematopoietic stem cell niche and regulator of primitive hematopoietic progenitor cells. *Blood, 106*, 1232–1239.

Nishita, M., Hashimoto, M. K., Ogata, S., et al. (2000). Interaction between Wnt and TGF-beta signalling pathways during formation of Spemann's organizer. *Nature, 403*, 781–785.

Noctor, S. C., Martínez-Cerdeño, V., Ivic, L., et al. (2004). Cortical neurons arise in symmetric and asymmetric division zones and migrate through specific phases. *Nature Neuroscience, 7*, 136–144.

Nyeng, P., Norgaard, G. A., Kobberup, S., et al. (2008). FGF10 maintains distal lung bud epithelium and excessive signaling leads to progenitor state arrest, distalization, and goblet cell metaplasia. *BMC Developmental Biology, 8*, 2.

Ohishi, K., Varnum-Finney, B., & Bernstein, I. D. (2002). Delta-1 enhances marrow and thymus repopulating ability of human CD34(+)CD38(−) cord blood cells. *The Journal of Clinical Investigation, 110*, 1165–1174.

Ohlstein, B., & Spradling, A. (2006). The adult Drosophila posterior midgut is maintained by pluripotent stem cells. *Nature, 439*(7075), 470–474.

Ohlstein, B., & Spradling, A. (2007). Multipotent Drosophila intestinal stem cells specify daughter cell fates by differential notch signaling. *Science, 315*(5814), 988–992.

Okada, S., Nakauchi, H., Nagayoshi, K., et al. (1992). In vivo and in vitro stem cell function of c-kit- and Sca-1-positive murine hematopoietic cells. *Blood, 80*, 3044–3050.

Okubo, T., Knoepfler, P. S., Eisenman, R. N., et al. (2005). N-myc plays an essential role during lung development as a dosage-sensitive regulator of progenitor cell proliferation and differentiation. *Development, 132*, 1363–1374.

Oliver, J. R., Kushwah, R., Wu, J., et al. (2011). Elf3 plays a role in regulating bronchiolar epithelial repair kinetics following Clara cell-specific injury. *Laboratory Investigation, 91*, 1514–1529.

Omran, A., Elimam, D., & Yin, F. (2013). MicroRNAs: New insights into chronic childhood diseases. *BioMed Research International, 2013*, 291826.

Osawa, M., Hanada, K., Hamada, H., et al. (1996). Long-term lymphohematopoietic reconstitution by a single CD34-low/negative hematopoietic stem cell. *Science, 273*, 242–245.

Overeem, A. W., Bryant, D. M., & van IJzendoorn, S. C. (2015). Mechanisms of apical-basal axis orientation and epithelial lumen positioning. *Trends in Cell Biology, 25*, 476–485.

Pandit, K. V., & Milosevic, J. (2015). MicroRNA regulatory networks in idiopathic pulmonary fibrosis. *Biochemistry and Cell Biology, 93*(2), 129–137.

Pardo-Saganta, A., Law, B. M., Tata, P. R., et al. (2015). Injury induces direct lineage segregation of functionally distinct airway basal stem/progenitor cell subpopulations. *Cell Stem Cell, 16*, 184–197.

Pardo-Saganta, A., Tata, P. R., Law, B., et al. (2015b). Parent stem cells can serve as niches for their daughter cells. *Nature, 523*, 597–601.

Parmar, K., Mauch, P., Vergilio, J. A., et al. (2007). Distribution of hematopoietic stem cells in the bone marrow according to regional hypoxia. *Proceedings of the National Academy of Sciences of the United States of America, 104*, 5431–5436.

Passier, R., Orlova, V., & Mummery, C. (2016). Complex tissue and disease modeling using hiPSCs. *Cell Stem Cell, 18*, 309–321.

Pece, S., Confalonieri, S. R., Romano, P., et al. (2011). NUMB-ing down cancer by more than just a NOTCH. *Biochimica et Biophysica Acta, 1815*, 26–43.

Peng, Y., & Axelrod, J. D. (2012). Asymmetric protein localization in planar cell polarity: Mechanisms, puzzles and challenges. *Current Topics in Developmental Biology, 101*, 33–53.

Peng, T., Frank, D. B., Kadzik, R. S., et al. (2015). Hedgehog actively maintains adult lung quiescence and regulates repair and regeneration. *Nature, 526*, 578–582.

Pepicelli, C. V., Lewis, P. M., & McMahon, A. P. (1998). Sonic hedgehog regulates branching morphogenesis in the mammalian lung. *Current Biology, 8*, 1083–1086.

Perl, A. K., Wert, S. E., Loudy, D. E., et al. (2005). Conditional recombination reveals distinct subsets of epithelial cells in trachea, bronchi, and alveoli. *American Journal of Respiratory Cell and Molecular Biology, 33*, v455–v462.

Pestina, T.I., Cleveland, J., Yang, C., Zambetti, G., & Jackson C (2001). Mpl ligand prevents lethal myelosuppression by inhibiting p53-dependent apoptosis. Blood, 98, 2084–2090.

Plantier, L., Marchand-Adam, S., Antico Arciuch, V. G., et al. (2007). Keratinocyte growth factor protects against elastase-induced pulmonary emphysema in mice. *American Journal of Physiology. Lung Cellular and Molecular Physiology, 293*, L1230–L1239.

Plopper, C., St George, J., Cardoso, W., et al. (1992). Development of airway epithelium. Patterns of expression for markers of differentiation. *Chest, 101*, 2S–5S.

Popova, A. P., Bentley, J. K., Anyanwu, A. C., et al. (2012). Glycogen synthase kinase-3ß/ß-catenin signaling regulates neonatal lung mesenchymal stromal cell myofibroblastic differentiation. *American Journal of Physiology. Lung Cellular and Molecular Physiology, 303*, L439–L438.

Quantius, J., Schmoldt, C., Vazquez-Armendariz, A. I., et al. (2016). Influenza virus infects epithelial stem/progenitor cells of the distal lung: Impact on Fgfr2b-driven epithelial repair. *PLoS Pathogens, 12*, e1005544.

Que, J., Okubo, T., Goldenring, J. R., et al. (2007). Multiple dose-dependent roles for Sox2 in the patterning and differentiation of anterior foregut endoderm. *Development, 134*, 2521–2531.

Que, J., Wilm, B., Hasegawa, H., et al. (2008). Mesothelium contributes to vascular smooth muscle and mesenchyme during lung development. *Proceedings of the National Academy of Sciences of the United States of America, 105*, 16626–16630.

Quiat, D., & Olson, E. N. (2013). MicroRNAs in cardiovascular disease: From pathogenesis to prevention and treatment. *The Journal of Clinical Investigation, 123*(1), 11–18.

Radtke, F., Wilson, A., Mancini, S. J., et al. (2004). Notch regulation of lymphocyte development and function. *Nature Immunology, 5*, 247–253.

Rafii, S., Cao, Z., Lis, R., et al. (2015). Platelet-derived SDF-1 primes the pulmonary capillary vascular niche to drive lung alveolar regeneration. *Nature Cell Biology, 17*, 123–136.

Ramasamy, S. K., Mailleux, A. A., Gupte, V. V., et al. (2007). Fgf10 dosage is critical for the amplification of epithelial cell progenitors and for the formation of multiple mesenchymal lineages during lung development. *Developmental Biology, 307*, 237–247.

Rankin, S. A., & Zorn, A. M. (2014). Gene regulatory networks governing lung specification. *Journal of Cellular Biochemistry, 115*(8), 1343–1350.

Rasin, M. R., Gazula, V. R., Breunig, J. J., Kwan, K. Y., Johnson, M. B., Liu-Chen, S., Li, H. S., Jan, L. Y., Jan, Y. N., Rakic, P., & Sestan, N. (2007). Numb and Numbl are required for maintenance of cadherin-based adhesion and polarity of neural progenitors. *Nature Neuroscience, 10*(7), 819–827.

Rawlins, E. L. (2008). Lung epithelial progenitor cells: Lessons from development. *Proceedings of the American Thoracic Society, 5*, 675–681.

Rawlins, E. L. (2015). Stem cells: Emergency back-up for lung repair. *Nature, 517*, 556–557.

Rawlins, E. L., & Hogan, B. L. (2006). Epithelial stem cells of the lung: Privileged few or opportunities for many? *Development, 133*, 2455–2465.

Rawlins, E. L., Ostrowski, L. E., Randell, S. H., et al. (2007). Lung development and repair: Contribution of the ciliated lineage. *Proceedings of the National Academy of Sciences of the United States of America, 104*, 410–417.

Rawlins, E. L., Clark, C. P., Xue, Y., et al. (2009a). The Id2+ distal tip lung epithelium contains individual multipotent embryonic progenitor cells. *Development, 136*, 3741–3745.

Rawlins, E. L., Okubo, T., Xue, Y., et al. (2009b). The role of Scgb1a1+ Clara cells in the long-term maintenance and repair of lung airway, but not alveolar, epithelium. *Cell Stem Cell, 4*, 525–534.

Ray, P., Devaux, Y., Stolz, D. B., et al. (2003). Inducible expression of keratinocyte growth factor (KGF) in mice inhibits lung epithelial cell death induced by hyperoxia. *Proceedings of the National Academy of Sciences of the United States of America, 100*, 6098–6103.

Ray, S., Chiba, N., Yao, C., et al. (2016). Rare SOX2(+) airway progenitor cells generate KRT5(+) cells that repopulate damaged alveolar parenchyma following influenza virus infection. *Stem Cell Reports, 7*, 817–825.

Reddy, R., Buckley, S., Doerken, M., et al. (2004). Isolation of a putative progenitor subpopulation of alveolar epithelial type 2 cells. *American Journal of Physiology. Lung Cellular and Molecular Physiology, 286*, L658–L657.

Reinhart, B. J., Slack, F. J., Basson, M., Pasquinelli, A. E., Bettinger, J. C., Rougvie, A. E., Horvitz, H. R., & Ruvkun, G. (2000). The 21-nucleotide let-7 RNA regulates developmental timing in Caenorhabditis elegans. *Nature, 403*(6772), 901–906.

Reya, T., Duncan, A. W., Ailles, L., et al. (2003). A role for Wnt signalling in self-renewal of haematopoietic stem cells. *Nature, 423*, 409–414.

Reynolds, S. D., Giangreco, A., Power, J. H., et al. (2000). Neuroepithelial bodies of pulmonary airways serve as a reservoir of progenitor cells capable of epithelial regeneration. *The American Journal of Pathology, 156*, 269–278.

Reynolds, S. D., Giangreco, A., Hong, K. U., et al. (2004). Airway injury in lung disease pathophysiology: Selective depletion of airway stem and progenitor cell pools potentiates lung inflammation and alveolar dysfunction. *American Journal of Physiology. Lung Cellular and Molecular Physiology, 287*, L1256–L1265.

Rizzo, D. C. (2016). *Fundamentals of anatomy and physiology* (4th ed.532 pages). GENGAGE Learning Publisher.

Rock, J. R., & Hogan, B. L. (2011). Epithelial progenitor cells in lung development, maintenance, repair, and disease. *Annual Review of Cell and Developmental Biology, 27*, 493–512.

Rock, J. R., Onaitis, M. W., Rawlins, E. L., et al. (2009). Basal cells as stem cells of the mouse trachea and human airway epithelium. *Proceedings of the National Academy of Sciences of the United States of America, 106*, 12771–12775.

Rock, J. R., Randell, S. H., & Hogan, B. L. (2010). Airway basal stem cells: A perspective on their roles in epithelial homeostasis and remodeling. *Disease Models & Mechanisms, 3*, 545–556.

Rock, J. R., Gao, X., Xue, Y., et al. (2011). Notch-dependent differentiation of adult airway basal stem cells. *Cell Stem Cell, 8*, 639–648.

Roignot, J., Peng, X., & Mostov, K. (2013). Polarity in mammalian epithelial morphogenesis. *Cold Spring Harbor Perspectives in Biology, 5*, a013789.

Roper, J. M1., Mazzatti, D. J., Watkins, R. H., Maniscalco, W. M., Keng, P. C., & O'Reilly, M. A. (2004). In vivo exposure to hyperoxia induces DNA damage in a population of alveolar type II epithelial cells. *American Journal of Physiology. Lung Cellular and Molecular Physiology, 286*(5), L1045–L1054.

Rutter, M., Wang, J., Huang, Z., et al. (2010). Gli2 influences proliferation in the developing lung through regulation of cyclin expression. *American Journal of Respiratory Cell and Molecular Biology, 42*, 615–625.

Sanada, K., & Tsai, L. H. (2005). G protein betagamma subunits and AGS3 control spindle orientation and asymmetric cell fate of cerebral cortical progenitors. *Cell, 122*(1), 119–131.

Sato, T., Vries, R. G., Snippert, H. J., et al. (2009). Single Lgr5 stem cells build crypt-villus structures in vitro without a mesenchymal niche. *Nature, 459*, 262–265.

Sato, T., van Es, J. H., Snippert, H. J., et al. (2011a). Paneth cells constitute the niche for Lgr5 stem cells in intestinal crypts. *Nature, 469*, 415–418.

Sato, T., Stange, D. E., Ferrante, M., et al. (2011b). Long-term expansion of epithelial organoids from human Colon, adenoma, adenocarcinoma, and Barrett's epithelium. *Gastroenterology, 141*, 1762–1772.

Sayed, D., & Abdellatif, M. (2011). MicroRNAs in development and disease. *Physiological Reviews, 91*(3), 827–887.

Schittny, J. C. (2017). Development of the lung. *Cell and Tissue Research, 367*(3), 427–444.

Schofield, R. (1978). The relationship between the spleen colony-forming cell and the haemopoietic stem cell. *Blood Cells, 4*, 7–25.

Seery, J. P., & Watt, F. M. (2000). Asymmetric stem-cell divisions define the architecture of human oesophageal epithelium. *Current Biology, 10*(22), 1447–1450.

Seita, J., & Weissman, I. L. (2010). Hematopoietic stem cell: Self-renewal versus differentiation. *Wiley Interdisciplinary Reviews. Systems Biology and Medicine, 2*, 640–653.

Seita, J., Ema, H., Ooehara, J., et al. (2007). Lnk negatively regulates self-renewal of hematopoietic stem cells by modifying thrombopoietin-mediated signal transduction. *Proceedings of the National Academy of Sciences of the United States of America, 104*, 2349–2354.

Serls, A. E., Doherty, S., Parvatiyar, P., et al. (2005). Different thresholds of fibroblast growth factors pattern the ventral foregut into liver and lung. *Development, 132*, 35–47.

Senoo, M1., Pinto, F., Crum, C. P., & McKeon, F. (2007). p63 Is essential for the proliferative potential of stem cells in stratified epithelia. *Cell, 129*(3), 523–536.

Seth, R., Shum, L., Wu, F., et al. (1993). Role of epidermal growth factor expression in early mouse embryo lung branching morphogenesis in culture: Antisense oligodeoxynucleotide inhibitory strategy. *Developmental Biology, 158*, 555–559.

Seymour, P. A., Freude, K. K., Tran, M. N., et al. (2007). SOX9 is required for maintenance of the pancreatic progenitor cell pool. *Proceedings of the National Academy of Sciences of the United States of America, 104*, 1865–1870.

Sgantzis, N., Yiakouvaki, A., Remboutsika, E., et al. (2011). HuR controls lung branching morphogenesis and mesenchymal FGF networks. *Developmental Biology, 354*, 267–279.

Shan, L., Subramaniam, M., Emanuel, R. L., et al. (2008). Centrifugal migration of mesenchymal cells in embryonic lung. *Developmental Dynamics, 237*, 750–757.

Shi, W., Xu, J., & Warburton, D. (2009). Development, repair and fibrosis: What is common and why it matters. *Respirology, 14*, 656–655.

Shin, J., Poling, J., Park, H. C., & Appel, B. (2007). Notch signaling regulates neural precursor allocation and binary neuronal fate decisions in zebrafish. *Development, 134*(10), 1911–1920.

Shinin, V., Gayraud-Morel, B., Gomes, D., & Tajbakhsh, S. (2006). Asymmetric division and cosegregation of template DNA strands in adult muscle satellite cells. *Nature Cell Biology, 8*, 677–687.

Shu, W., Guttentag, S., Wang, Z., et al. (2005). Wnt/beta-catenin signaling acts upstream of N-myc, BMP4, and FGF signaling to regulate proximal-distal patterning in the lung. *Developmental Biology, 283*, 226–239.

Shu, W., Lu, M. M., Zhang, Y., et al. (2007). Foxp2 and Foxp1 cooperatively regulate lung and esophagus development. *Development, 134*, 1991–2000.

Sitnicka, E., Ruscetti, F. W., Priestley, G. V., et al. (1996). Transforming growth factor beta 1 directly and reversibly inhibits the initial cell divisions of long-term repopulating hematopoietic stem cells. *Blood, 88*, 82–88.

Smart, I. H. (1970). Variation in the plane of cell cleavage during the process of stratification in the mouse epidermis. *The British Journal of Dermatology, 82*(3), 276–282.

Smith, C. Al., Lau, K. M., Rahmani, Z., Dho, S. E., Brothers, G., She, Y. M., Berry, D. M., Bonneil, E., Thibault, P., Schweisguth, F., Le Borgne, R., & McGlade, C. J. (2007). aPKC-mediated phosphorylation regulates asymmetric membrane localization of the cell fate determinant Numb. *The EMBO Journal, 26*(2), 468–480.

Snitow, M., Lu, M., Cheng, L., et al. (2016). Ezh2 restricts the smooth muscle lineage during mouse lung mesothelial development. *Development, 143*, 3733–3741.

Solar, G. P., Kerr, W. G., Zeigler, F. C., et al. (1998). Role of c-mpl in early hematopoiesis. *Blood, 92*, 4–10.

Sountoulidis, A., Stavropoulos, A., Giaglis, S., et al. (2012). Activation of the canonical bone morphogenetic protein (BMP) pathway during lung morphogenesis and adult lung tissue repair. *PLoS One, 7*, e41460.

Spangrude, G. J., Heimfeld, S., & Weissman, I. L. (1988). Purification and characterization of mouse hematopoietic stem cells. *Science, 241*, 58–62.

Spurlin, J. W., III, & Nelson, C. M. (2017). Building branched tissue structures: From single cell guidance to coordinated construction. *Philosophical Transactions of the Royal Society B, 372*, 20150527.

Stabler, C. T., & Morrisey, E. E. (2017). Developmental pathways in lung regeneration. *Cell and Tissue Research, 367*, 677.

Stevens, T., Phan, S., Frid, M. G., et al. (2008). Lung vascular cell heterogeneity: Endothelium, smooth muscle, and fibroblasts. *Proceedings of the American Thoracic Society, 5*, 783–791.

Stier, S., Cheng, T., Dombkowski, D., et al. (2002). Notch1 activation increases hematopoietic stem cell self-renewal in vivo and favors lymphoid over myeloid lineage outcome. *Blood, 99*, 2369–2378.

Stier, S., Ko, Y., Forkert, R., et al. (2005). Osteopontin is a hematopoietic stem cell niche component that negatively regulates stem cell pool size. *The Journal of Experimental Medicine, 201*, 1781–1791.

Stoltz, J.-F., de Isla, N., Li, Y. P., et al. (2015). Stem cells and regenerative medicine: Myth or reality of the 21th century. *Stem Cells International, 2015*, 734731.

Sucre, J. M. S., Wilkinson, D., Vijayaraj, P., et al. (2016). A three-dimensional human model of the fibroblast activation that accompanies bronchopulmonary dysplasia identifies Notch-mediated pathophysiology. *American Journal of Physiology. Lung Cellular and Molecular Physiology, 310*, L889–L898.

Sun, R., Zhou, Q., Ye, X., Takahata, T., Ishiguro, A., Kijima, H., Nukiwa, T., & Saijo, Y. (2013). A change in the number of CCSP (pos)/SPC(pos) cells in mouse lung during development, growth, and repair. *Respiratory Investigation, 51*, 229–240.

Suzuki, A., & Ohno, S. (2006). The PAR-aPKC system: Lessons in polarity. *Journal of Cell Science, 119*.(Pt 6, 979–987.

Suzuki, A., Raya, A., Kawakami, Y., et al. (2006). Nanog binds to Smad1 and blocks bone morphogenetic protein-induced differentiation of embryonic stem cells. *Proceedings of the National Academy of Sciences of the United States of America, 103*, 10294–10299.

Tadokoro, T., Wang, Y., Barak, L. S., et al. (2014). IL-6/STAT3 promotes regeneration of airway ciliated cells from basal stem cells. *Proceedings of the National Academy of Sciences of the United States of America, 111*, E3641–E3649.

Tadokoro, T., Gao, X., Hong, C. C., et al. (2016). BMP signaling and cellular dynamics during regeneration of airway epithelium from basal progenitors. *Development, 143*, 764–773.

Takahashi, K., & Yamanaka, S. (2006). Induction of pluripotent stem cells from mouse embryonic and adult fibroblast cultures by defined factors. *Cell, 126*, 663–676.

Takahashi, Y., Izumi, Y., Kohno, M., et al. (2010). Thyroid transcription factor-1 influences the early phase of compensatory lung growth in adult mice. *American Journal of Respiratory and Critical Care Medicine, 181*, 1397–1406.

Takaki, S., Sauer, K., Iritani, B. M., et al. (2000). Control of B cell production by the adaptor protein lnk: Definition of a conserved family of signal-modulating proteins. *Immunity, 13*, 599–609.

Takasato, M., Er, P. X., Chiu, H. S., et al. (2016). Kidney organoids from human iPS cells contain multiple lineages and model human nephrogenesis. *Nature, 536*, 238–238.

Tata, P. R., & Rajagopal, J. (2016). Regulatory circuits and bi-directional signaling between stem cells and their progeny. *Cell Stem Cell, 19*, 686–689.

Tata, P. R., & Rajagopal, J. (2017). Plasticity in the lung: Making and breaking cell identity. *Development, 144*, 755–766.

Tata, P. R., Mou, H., Pardo-Saganta, A., et al. (2013). Dedifferentiation of committed epithelial cells into stem cells in vivo. *Nature, 503*, 218–223.

Tefft, D., Lee, M., Smith, S., et al. (2002). mSprouty2 inhibits FGF10-activated MAP kinase by differentially binding to upstream target proteins. *American Journal of Physiology. Lung Cellular and Molecular Physiology, 283*, L700–L706.

Tefft, D., De Langhe, S. P., Del Moral, P. M., et al. (2005). A novel function for the protein tyrosine phosphatase Shp2 during lung branching morphogenesis. *Developmental Biology, 282*, 422–431.

Teisanu, R. M., Chen, H., Matsumoto, K., et al. (2011). Functional analysis of two distinct bronchiolar progenitors during lung injury and repair. *American Journal of Respiratory Cell and Molecular Biology, 44*, 794–803.

Tepass, U. (2012). The apical polarity protein network in Drosophila epithelial cells: Regulation of polarity, junctions, morphogenesis, cell growth, and survival. *Annual Review of Cell and Developmental Biology, 28*, 655–685.

Tian, Y., Zhang, Y., Hurd, L., Hannenhalli, S., Liu, F., Lu, M. M., & Morrisey, E. E. (2011). Regulation of lung endoderm progenitor cell behavior by miR302/367. *Development, 138*, 1235–1245.

Till, J. E., McCulloch, E. A., & Siminovitch, L. (1964). A stochastic model of stem cell proliferation, based on the growth of spleen colony-forming cells. *Proceedings of the National Academy of Sciences of the United States of America, 51*, 29–36.

Toledano, H., & Jones, D. L. (2008). *Mechanisms regulating stem cell polarity and the specification of asymmetric divisions. StemBook [Internet].* Cambridge (MA): Harvard Stem Cell Institute 2008-2009 31.

Tong, W., Zhang, J., & Lodish, H. F. (2005). Lnk inhibits erythropoiesis and Epo-dependent JAK2 activation and downstream signaling pathways. *Blood, 105,* 4604–4612.

Treutlein, B., Brownfield, D. G., Wu, A. R., et al. (2014). Reconstructing lineage hierarchies of the distal lung epithelium using single-cell RNA-seq. *Nature, 509,* 371–375.

Trowbridge, J. J., Xenocostas, A., Moon, R. T., et al. (2006). Glycogen synthase kinase-3 is an in vivo regulator of hematopoietic stem cell repopulation. *Nature Medicine, 12,* 89–98.

Tsao, P. N., Wei, S. C., Wu, M. F., et al. (2011). Notch signaling prevents mucous metaplasia in mouse conducting airways during postnatal development. *Development, 138,* 3533–3543.

Tsao, P. N., Matsuoka, C., Wei, S. C., et al. (2016). Epithelial Notch signaling regulates lung alveolar morphogenesis and airway epithelial integrity. *Proceedings of the National Academy of Sciences of the United States of America, 113,* 8242–8247.

Tuder, R. M., & Yun, J. H. (2008). Vascular endothelial growth factor the lung: friend or foe. *Current Opinion in Pharmacology, 8*(3), 255–260 PMC. Web. 20 Feb. 2018.

Tuncay, H., & Ebnet, K. (2016). Cell adhesion molecule control of planar spindle orientation. *Cellular and Molecular Life Sciences, 73,* 1195–1207.

Varnum-Finney, B., Xu, L., Brashem-Stein, C., et al. (2000). Pluripotent, cytokine-dependent, hematopoietic stem cells are immortalized by constitutive Notch1 signaling. *Nature Medicine, 6,* 1278–1281.

Varnum-Finney, B., Brashem-Stein, C., & Bernstein, I. D. (2003). Combined effects of Notch signaling and cytokines induce a multiple log increase in precursors with lymphoid and myeloid reconstituting ability. *Blood, 101,* 1784–1789.

Vas, V., Szilagyi, L., Paloczi, K., et al. (2004). Soluble Jagged-1 is able to inhibit the function of its multivalent form to induce hematopoietic stem cell self-renewal in a surrogate in vitro assay. *Journal of Leukocyte Biology, 75,* 714–720.

Vaughan, A. E., Brumwell, A. N., Xi, Y., et al. (2015). Lineage-negative progenitors mobilize to regenerate lung epithelium after major injury. *Nature, 517,* 621–625.

Velazquez, L., Cheng, A. M., Fleming, H. E., et al. (2002). Cytokine signaling and hematopoietic homeostasis are disrupted in Lnk-deficient mice. *The Journal of Experimental Medicine, 195,* 1599–1611.

Volckaert, T., Campbell, A., Dill, E., et al. (2013). Localized Fgf10 expression is not required for lung branching morphogenesis but prevents differentiation of epithelial progenitors. *Development, 140,* 3731–3742.

Vorhagen, S., & Niessen, C. M. (2014). Mammalian aPKC/Par polarity complex mediated regulation of epithelial division orientation and cell fate. *Experimental Cell Research, 328,* 296–302.

Wan, H., Dingle, S., Xu, Y., et al. (2005). Compensatory roles of Foxa1 and Foxa2 during lung morphogenesis. *The Journal of Biological Chemistry, 280,* 13809–13816.

Wang, C., Chang, K. C., Somers, G., et al. (2009). Protein phosphatase 2A regulates self-renewal of Drosophila neural stem cells. *Development, 136,* 2287–2296.

Wang, Y., Tian, Y., Morley, M. P., et al. (2013). Development and regeneration of Sox2+ endoderm progenitors are regulated by a Hdac1/2- Bmp4/Rb1 regulatory pathway. *Developmental Cell, 24,* 345–358.

Wang, X., Wang, Y., Snitow, M. E., et al. (2016). Expression of histone deacetylase 3 instructs alveolar type I cell differentiation by regulating a Wnt signaling niche in the lung. *Developmental Biology, 414,* 161–169.

Warburton, D. (2008). Developmental biology: Order in the lung. *Nature, 453,* 73–75.

Warburton, D., Schwarz, M., Tefft, D., et al. (2000). The molecular basis of lung morphogenesis. *Mech Dev, 92,* 55–81.

Warburton, D., Perin, L., Defilippo, R., et al. (2008). Stem/progenitor cells in lung development, injury repair, and regeneration. *Proceedings of the American Thoracic Society, 5,* 703–706.

Warburton, D., El-Hashash, A., Carraro, G., et al. (2010). Lung organogenesis. *Current Topics in Developmental Biology, 90,* 73–158.

Watson, C. L., Mahe, M. M., Múnera, J., et al. (2014). An in vivo model of human small intestine using pluripotent stem cells. *Nature Medicine, 20,* 1310–1314.

Watson, J. K., Rulands, S., Wilkinson, A. C., et al. (2015). Clonal dynamics reveal two distinct populations of basal cells in slow-turnover airway epithelium. *Cell Reports, 12,* 90–101.

Weaver, M., Yingling, J. M., Dunn, N. R., et al. (1999). Bmp signaling regulates proximal-distal differentiation of endoderm in mouse lung development. *Development, 126,* 4005–4015.

Weibel, E. R. (2015). On the tricks alveolar epithelial cells play to make a good lung. *Am J Respir Crit Care Med, 191*(5), 504–513.

Whitsett, J. A., Haitchi, H. M., & Maeda, Y. (2011). Intersections between pulmonary development and disease. *American Journal of Respiratory and Critical Care Medicine, 184,* 401–406.

Wilkinson, D. C., Alva-Ornelas, J. A., Sucre, J. M. S., et al. (2016). Development of a three-dimensional bioengineering technology to generate lung tissue for personalized disease modeling. *Stem Cells Translational Medicine, 6*(2), 622–633.

Willert, K., Brown, J. D., Danenberg, E., et al. (2003). Wnt proteins are lipid-modified and can act as stem cell growth factors. *Nature, 423,* 448–452.

Wilson, A., & Trumpp, A. (2006). Bone-marrow haematopoietic-stem-cell niches. *Nature Reviews. Immunology, 6,* 93–106.

Wodarz, A. (2002). Establishing cell polarity in development. *Nature Cell Biology, 4,* E39–E44.

Wong, A. P., Bear, C. E., Chin, S., et al. (2012). Directed differentiation of human pluripotent stem cells into mature airway epithelia expressing functional CFTR protein. *Nature Biotechnology, 30,* 876–882.

Wong, A. P., Chin, S., Xia, S., et al. (2015). Efficient generation of functional CFTR-expressing airway epithelial cells from human pluripotent stem cells. *Nature Protocols, 10,* 363–381.

Woods, D. F., Wu, J. W., & Bryant, P. J. (1997). Localization of proteins to the apico-lateral junctions of Drosophila epithelia. *Developmental Genetics, 20,* 111–118.

Wu, M., Kwon, H. Y., Rattis, F., Blum, J., Zhao, C., Ashkenazi, R., Jackson, T. L., Gaiano, N., Oliver, T., & Reya, T. (2007). Imaging hematopoietic precursor division in real time. *Cell Stem Cell, 1*(5), 541–554.

Yamashita, Y. M. (2009). Regulation of asymmetric stem cell division: Spindle orientation and the centrosome. *Frontiers in Bioscience (Landmark Ed)., 14,* 3003–3011.

Yamashita, Y. M., Yuan, H., Cheng, J., et al. (2010). Polarity in stem cell division: Asymmetric stem cell division in tissue homeostasis. *Cold Spring Harbor Perspectives in Biology, 2,* a001313.

Yan, B., Omar, F. M., Das, K., et al. (2008). Characterization of Numb expression in astrocytomas. *Neuropathology, 28,* 479–484.

Yang, A., Schweitzer, R., Sun, D., Kaghad, M., Walker, N., Bronson, R. T., Tabin, C., Sharpe, A., Caput, D., Crum, C., & McKeon, F. (1999). p63 is essential for regenerative proliferation in limb, craniofacial and epithelial development. *Nature, 398*(6729), 714–718.

Yang, L., Bryder, D., Adolfsson, J., et al. (2005). Identification of Lin(−)Sca1(+)kit(+)CD34(+)Flt3- short-term hematopoietic stem cells capable of rapidly reconstituting and rescuing myeloablated transplant recipients. *Blood, 105,* 2717–2723.

Yang, S., Ma, K., Geng, Z., et al. (2015). Oriented cell division: New roles in guiding skin wound repair and regeneration. *Bioscience Reports, 35,* 6.

Yin, Y., Wang, F., & Ornitz, D. M. (2011). Mesothelial- and epithelial-derived FGF9 have distinct functions in the regulation of lung development. *Development, 138,* 3169–3177.

Yin, X., Farin, H. F., van Es, J. H., et al. (2014). Niche-independent high purity cultures of Lgr5+ intestinal stem cells and their progeny. *Nature Methods, 11,* 106–112.

You, Y., Richer, E. J., Huang, T., & Brody, S. L. (2002). Growth and differentiation of mouse tracheal epithelial cells: Selection of a proliferative population. *American Journal of Physiology. Lung Cellular and Molecular Physiology, 283,* L1315–L1321.

Zhang, Y., Goss, A. M., Cohen, E. D., et al. (2008). A Gata6-Wnt pathway required for epithelial stem cell development and airway regeneration. *Nature Genetics, 40,* 862–870.

Zhang, S., Zhou, X., Chen, T., et al. (2014). Single primary fetal lung cells generate alveolar structures in vitro. *In Vitro Cellular & Developmental Biology. Animal, 50*, 87–93.

Zheng, J., Huynh, H., Umikawa, M., et al. (2011). Angiopoietin-like protein 3 supports the activity of hematopoietic stem cells in the bone marrow niche. *Blood, 117*, 470–479.

Zhou, Q., Law, A. C., Rajagopal, J., et al. (2007). A multipotent progenitor domain guides pancreatic organogenesis. *Developmental Cell, 13*, 103–114.

Zhong, W., Feder, J. N., Jiang, M. M., Jan, L. Y., & Jan, Y. N. (1996). Asymmetric localization of a mammalian numb homolog during mouse cortical neurogenesis. *Neuron, 17*(1), 43–53.

Zhong, W., Jiang, M. M., Weinmaster, G., Jan, L. Y., & Jan, Y. N. (1997). Differential expression of mammalian Numb, Numblike and Notch1 suggests distinct roles during mouse cortical neurogenesis. *Development, 124*(10), 1887–1897.

Zigman, M., Cayouette, M., Charalambous, C., Schleiffer, A., Hoeller, O., Dunican, D., McCudden, C. R., Firnberg, N., Barres, B. A., Siderovski, D. P., & Knoblich, J. A. (2005). Mammalian inscuteable regulates spindle orientation and cell fate in the developing retina. *Neuron, 48*(4), 539–545.

Zimdahl, B., Ito, T., Blevins, A., et al. (2014). Lis1 regulates asymmetric division in hematopoietic stem cells and in leukemia. *Nature Genetics, 46*, 245–252.

Zuo, W., Zhang, T., Wu, D. Z., et al. (2014). p63+Krt5+ distal airway stem cells are essential for lung regeneration. *Nature, 517*, 616–620.

Index

Printed in the United States
By Bookmasters